Tourism and Brexit

TOURISM AND CULTURAL CHANGE

Series Editors: Professor Mike Robinson, *Ironbridge International Institute for Cultural Heritage, University of Birmingham, UK* and Professor Alison Phipps, *University of Glasgow, Scotland, UK*

Understanding tourism's relationships with culture(s) and vice versa, is of ever-increasing significance in a globalising world. TCC is a series of books that critically examine the complex and ever-changing relationship between tourism and culture(s). The series focuses on the ways that places, peoples, pasts and ways of life are increasingly shaped/transformed/created/packaged for touristic purposes. The series examines the ways tourism utilises/makes and re-makes cultural capital in its various guises (visual and performing arts, crafts, festivals, built heritage, cuisine, etc.) and the multifarious political, economic, social and ethical issues that are raised as a consequence. Theoretical explorations, research-informed analyses and detailed historical reviews from a variety of disciplinary perspectives are invited to consider such relationships.

All books in this series are externally peer-reviewed.

Full details of all the books in this series and of all our other publications can be found on http://www.channelviewpublications.com, or by writing to Channel View Publications, St Nicholas House, 31-34 High Street, Bristol BS1 2AW, UK.

TOURISM AND CULTURAL CHANGE: 56

Tourism and Brexit

Travel, Borders and Identity

Edited by
Hazel Andrews

CHANNEL VIEW PUBLICATIONS
Bristol • Blue Ridge Summit

DOI https://doi.org/10.21832/ANDREW7918
Library of Congress Cataloging in Publication Data
A catalog record for this book is available from the Library of Congress.
Names: Andrews, Hazel, editor.
Title: Tourism and Brexit: Travel, Borders and Identity/Edited by Hazel Andrews.
Description: Blue Ridge Summit: Channel View Publications, 2020. | Series: Tourism and Cultural Change: 56 | Includes bibliographical references and index. | Summary: "This book is the first to explore the relationship between tourism and Brexit from a social science perspective. Contributors from around the world use international examples to examine three entwined themes integral to tourism: travel, borders and identity. It will be useful for students and researchers in tourism, migration and European studies"—Provided by publisher.
Identifiers: LCCN 2020024866 (print) | LCCN 2020024867 (ebook) | ISBN 9781845417901 (Paperback) | ISBN 9781845417918 (Hardback) | ISBN 9781845417925 (PDF) | ISBN 9781845417932 (ePub) | ISBN 9781845417949 (Kindle Edition)
Subjects: LCSH: Tourism—Great Britain. | European Union—Great Britain. | Tourism—Social aspects—Great Britain.
Classification: LCC G155.G7 T654 2020 (print) | LCC G155.G7 (ebook) | DDC 338.4/79141—dc23 LC record available at https://lccn.loc.gov/2020024866
LC ebook record available at https://lccn.loc.gov/2020024867

British Library Cataloguing in Publication Data
A catalogue entry for this book is available from the British Library.

ISBN-13: 978-1-84541-791-8 (hbk)
ISBN-13: 978-1-84541-790-1 (pbk)

Channel View Publications
UK: St Nicholas House, 31-34 High Street, Bristol BS1 2AW, UK.
USA: NBN, Blue Ridge Summit, PA, USA.

Website: www.channelviewpublications.com
Twitter: Channel_View
Facebook: https://www.facebook.com/channelviewpublications
Blog: www.channelviewpublications.wordpress.com

Copyright © 2021 Hazel Andrews and the authors of individual chapters

All rights reserved. No part of this work may be reproduced in any form or by any means without permission in writing from the publisher.

The policy of Multilingual Matters/Channel View Publications is to use papers that are natural, renewable and recyclable products, made from wood grown in sustainable forests. In the manufacturing process of our books, and to further support our policy, preference is given to printers that have FSC and PEFC Chain of Custody certification. The FSC and/or PEFC logos will appear on those books where full certification has been granted to the printer concerned.

Typeset by Deanta Global Publishing Services, Chennai, India

Contents

Figures and Tables   vii

Contributors   viii

Foreword   xiii
Tom Selwyn

1   The Magic and Liminality of Tourism and Brexit   1
     Hazel Andrews

2   'Travel broadens the mind and enriches the soul':
     Exploring the Significance of Tourism and Brexit
     on the Spiritual Dimension of European Integration   18
     Alexandra Pimor

3   The Order of Things: Brexit and the Challenge of Identity   34
     Catherine Palmer

4   Uses of the Past: Heritage, Tourism and the Challenges of
     (Re)Producing Contemporary National Identities in England   48
     Vivian B. Gornik

5   Royal Events and Tourism in the Post-Brexit Era   66
     Jennifer Frost and Warwick Frost

6   Brexit and Tourism in Central and Eastern Europe:
     The Case of Poland   80
     Sabina Owsianowska and Magdalena Banaszkiewicz

7   From Duty Free to Benidorm: British Tourists in Spain
     in an Age of Brexit   96
     Mark Casey

8   Taking Back Control: The Freedom of the Holiday   110
     Hazel Andrews

9	Divisions and Schisms in the Party Space *Anne Storch and Nico Nassenstein*	125
10	Post-Brexit Tourism and the Commonwealth Reimagined *Marcus L. Stephenson and Shaun Goldfinch*	141
11	Brexit and the UK Overseas Territories: Tourism and the Reconstitution of Core–Periphery Identity *Maria Amoamo*	157
12	Associate EU Citizenship: A Panacea for Loss of Fundamental Rights, Mobility and European Identity Post-Brexit? *Victoria Hooton*	174
13	Brexit and Post-Globalisation Era: Walking into Unknown Geography *Reza Masoudi*	188

Coda 2020: COVID-19 Masks but Fails to Flatten Brexit   203
Hazel Andrews

Index   215

Figures and Tables

Figures

Figure 4.1	View looking out across the Tintagel Castle site (Photo credit: Vivian Gornik)	52
Figure 4.2	Tourists take photos of the sculpture *Gallos* which sits on the upper most level of the Tintagel Castle site (Photo credit: Vivian Gornik)	58
Figure 4.3	Souvenirs inside one of the two English Heritage gift shops at Tintagel Castle (Photo credit: Vivian Gornik)	59
Figure 5.1	Parade during the 1988 Trooping of the Colour (Source: Margaret Zallar, authors' collection)	68
Figure 5.2	Promotional poster for the *Tudors to Windsor: British Royal Portraits* Exhibition, Bendigo Art Gallery, Australia, depicting the coronation gown of Queen Elizabeth II (Source: Jennifer Frost)	72
Figure 9.1	Brexit as opportunity (Greser & Lenz, 2018)	126
Figure 9.2	Performed Britishness (Photo from Storch, 2018)	130
Figure 9.3	Visitors only (S'Arenal; photo Storch, 2017)	134
Figure 9.4	Brexit toilet paper purchased in Magaluf (Archive Storch)	138

Tables

Table 6.1	Tourism in Poland 2017	82
Table 11.1	Brexit and implications for the UKOTs	164
Table 11.2	Pitcairn EU-funded projects with tourism-related benefits	168

Contributors

Maria Amoamo is a Senior Research Fellow in the Department of Management at the University of Otago in New Zealand. She has a PhD in Tourism with particular interest in cultural and indigenous tourism development. Her current research examines aspects of cultural change and sustainable development in small island states and subnational island jurisdictions (SNIJ). She is particularly interested in how SNIJs can provide evidence of alternative and viable models of sustainable development, innovative governance, their dyadic relationship with the metropolitan or colonial power and its cultural, judicial and administrative legacy. From this, she has published on themes exploring social capital, tourism, symbolic communities, identity, sovereignty, resilience, vulnerability and sustainable livelihood strategies.

Hazel Andrews is Reader in Tourism, Culture & Society at Liverpool John Moores University. As a social anthropologist, Hazel is interested in issues of identity, selfhood and the body, principally in relation to tourism and travel. Hazel's PhD thesis was the first full-length ethnographic study of British charter tourists which involved periods of participant observation in the resorts of Palmanova and Magaluf on the Mediterranean island of Mallorca. Hazel has drawn on her fieldwork to publish journal articles and book chapters. She is the author/editor of seven books including the monograph *The British on Holiday* published by Channel View Publications in 2011 and the edited collection *Tourism and Violence* published by Ashgate in 2014.

Magdalena Banaszkiewicz is a Cultural Anthropologist. She is Lecturer in the Institute of Intercultural Studies at the Jagiellonian University and cooperates with Krakow University (Poland). She graduated both in Russian Studies and Cultural Studies. Her main research interests include cultural tourism and Central and Eastern European heritage (with a special focus on the difficult and dark aspects of it). She is an editorial board member of *Cultural Tourism*. Her doctoral dissertation 'The intercultural dialogue in tourism: The case of Polish and Russian relationships' was published in 2011. Recently, she co-edited a monograph *Anthropology of Tourism in Central and Eastern Europe: Bridging Worlds*.

Mark Casey is a Lecturer in Sociology at Newcastle University, UK. He has published on a number of areas including sexuality, gender, tourism and travel. Mark is co-editor with Thomas Thurnell-Read of *Men, Masculinities, Travel and Tourism* published by Palgrave in 2014.

Jennifer Frost is an Associate Professor in the Department of Management, Sport and Tourism at La Trobe University, Melbourne, Australia. Her research interests include travel narratives; the role of events in society; rural and regional regeneration through tourism and events; and health and wellness tourism. Her most recent co-authored research book is *Royal Events: Rituals, Innovations, Meanings* (Routledge, 2018). Jennifer is a co-editor of the Routledge Advances in Events Research series and was recognised in 2017 as an Emerging Scholar of Distinction by the International Academy for the Study of Tourism.

Warwick Frost is Professor of Tourism, Heritage and the Media in the Department of Management, Sport and Tourism at La Trobe University, Melbourne, Australia. His research interests include heritage, events, nature-based attractions and the interaction between media, popular culture and tourism. Warwick is a co-editor of the Routledge Advances in Events Research series and has co-edited six books and co-authored six research books. His most recent co-authored research book is *Royal Events: Rituals, Innovations, Meanings* (Routledge, 2018).

Shaun Goldfinch is the ANZSOG and WA Govt Chair in Public Policy and Administration at Curtin University, Western Australia. He has published five books and over 50 articles and chapters on public policy, public administration and development. His most recent book was *Prometheus Assessed* (co-authored with Kiyoshi Yamamoto). He has held positions at the University of Nottingham, the American University of Sharjah, the University of the South Pacific and the Universities of Otago and Canterbury. He has also held visiting positions in Japan, and consulted for governments around the world.

Vivian Gornik earned her PhD in Applied Anthropology from the University of South Florida in 2018. She also has an MA in Museum Studies and a BA in Anthropology from the University of Florida. Her research interests include exploring issues of heritage, museums, national identity and tourism. She conducted her dissertation fieldwork in England studying heritage tourism and national identity in Glastonbury and Tintagel in 2017. She has worked for museums including the Florida Holocaust Museum, the Sulphur Springs Museum and Heritage Center and the

Florida Museum of Natural History. In 2012, she interned with the Churches Conservation Trust in Bristol and it was that experience that solidified her interest in the British heritage sector.

Victoria Hooton is Lecturer in European Union Law at the University of Portsmouth. She researches in the highly topical area of EU free movement law and access to social security. This is an exciting and challenging part of EU and national law with socio-legal elements and provides much scope for Brexit-related research. Victoria has been a guest scholar for the National Trust, as well as undertaking a scholarship placement at the Max Planck Institute for European Legal History in Germany. She has presented papers at the University of Manchester, as well as at the Socio-Legal Studies Association. She was awarded her PhD by the University of Manchester.

Nico Nassenstein is Junior Professor for African Languages and Linguistics at the Johannes Gutenberg University Mainz. He holds a PhD from the University of Cologne and works mainly in the fields of sociolinguistics. His research interests include variation in Kiswahili, youth language and urban/rural language practices in Africa, language and identity and the linguistic aspects of violent conflict. He has recently developed an interest in the sociolinguistic aspects of language and (sex) tourism along the Kenyan coast. Several (co-edited) volumes/issues are currently in preparation, with a focus on linguistic creativity and secrecy (with A. Storch, A. Hollington and A. Aikhenvald, *Journal of Language and Culture*), youth languages (with A. Storch and A. Hollington, *The Mouth*) and swearing and cursing practices (with A. Storch, Mouton de Gruyter).

Reza Masoudi is a native south-western Iranian who lives in London. As an urbanist, his work focuses on the geography of crowds and protests, urban violence and extensive studies of religious rituals in public spaces in Iran and India. He has been a Research Fellow at the Max Planck Institute for the Study of Religious and Ethnic Diversity in Göttingen (Germany) and an Alexander von Humboldt Fellow at the Centre for Modern Oriental Studies (ZMO), Berlin. Reza is currently a Research Associate at SOAS, University of London. He received his PhD from The Bartlett Faculty of the Built Environment, UCL (2009).

Sabina Owsianowska is a Tourism Researcher and an Assistant Professor in the Department of the Theory of Leisure and Tourism at the University of Physical Education in Cracow (Poland) and cooperates with the Jagiellonian University. Her main research interests include anthropological aspects of tourism, cultural tourism, dissonant heritage in Central and Eastern Europe, tourism promotion and the embodiment of tourist experiences. She is an editorial board member of the scientific journal

Folia Turistica. Recently, she co-edited two monographs: *Anthropology of Tourism in Central and Eastern Europe: Bridging Worlds* (2018, with M. Banaszkiewicz) and *Anthropology of Tourism* (in Polish, 2017, with R. Winiarski).

Catherine Palmer is a Social Anthropologist in the Centre for Memory, Narrative and Histories, University of Brighton, UK. Her research focuses on identity, heritage and materiality; post-conflict/memorial landscapes; embodiment; the coast/seaside. She is a Fellow of the Royal Anthropological Institute and a member of the Association of Social Anthropologists. She is the joint series editor for the Routledge Advances in Tourism Anthropology book series (with Jo-Anne Lester) and the author of books and journal articles including the Routledge monograph *Being and Dwelling through Tourism: An Anthropological Perspective* and *Creating Heritage for Tourism* (with Jacqueline Tivers). Her edited (with Hazel Andrews) book for Routledge, *Tourism and Embodiment* was published in 2020.

Alexandra Pimor is Senior Law Lecturer at the LJMU School of Law, specialising in (EU and UK) public law, and she is co-author of *Unlocking EU Law* (with Tony Storey, Routledge 2018). Her studies focus on the intersection between law, governance and spirituality (not religion) at the European and international level, and on conceptualising the spiritual dimension in the field of European integration theories. More broadly, Alexandra is a scholar engaged in the advancement of nature rights and the reframing of legisprudential and jurisprudential orders based on the ecology of law, earth jurisprudence and ethics, and conscious governance paradigm.

Tom Selwyn is Leverhulme Emeritus Professorial Fellow at the Department of Anthropology and Sociology, SOAS, University of London. He is widely published in the field of the anthropology of tourism/pilgrimage/cultural heritage with regional interests in the Mediterranean, in particular Palestine/Israel. He directed/co-directed four major research and development projects in the Mediterranean region for the European Commission between 1995 and 2010, founded the MA in the Anthropology of Travel, Tourism and Pilgrimage (ATTP) at SOAS in 2010, co-convenes the SOAS ATTP summer school, was awarded the Lucy Mair medal by the Royal Anthropological Institute in 2009, and was Hon. Librarian of the RAI library for a decade. He is co-editor (with Nicola Frost) of the series Articulating Journeys: Festivals, Memorials, and Homecomings for Berghahn publishers, co-editor of *Travelling Towards Home* (2018) in that series, author of 'Tourism, Travel, and Pilgrimage' in Callan, H, (ed) (2018) *International Encyclopedia of Anthropology* (Wiley), 'Brexit, Grenfell, Windrush, and the mooring, un-mooring, and re-mooring of home' in *Ethnoscripts* (2019), 'The politics of Palestinian suffering, commemoration, and belonging', *JRAI* (2020), 'Why People Travel', *Independent Travel Commission* (2021).

Marcus L. Stephenson is a Professor of Tourism and Hospitality Management, and also Dean of the School of Hospitality at Sunway University (Malaysia). He has published extensively on the sociology of tourism, especially concerning issues of nationality, race, ethnicity, culture and religion. He is co-author of *Tourism and Citizenship: Rights, Freedoms and Responsibilities in the Global Order* (Routledge, 2014). He is also the co-editor of *International Tourism Development and the Gulf Cooperation Council States: Challenges and Opportunities* (Routledge, 2017).

Anne Storch is Professor of African Linguistics at the University of Cologne. Her principal research has been on the various languages of Nigeria (including Jukun and Maaka), on the Atlantic language region and on Western Nilotic (Southern Sudan and Uganda). Her work combines contributions on cultural and social contexts of languages, the semiotics of linguistic practices, epistemes and ontologies of colonial linguistics, as well as linguistic description. She has contributed to the analysis of registers and choices, language as social practice, ways of speaking and complex repertoires. Currently, she is interested in epistemic language, metalinguistics, noise and silence, as well as language use in complex settings, such as tourism. Her publications include *Secret Manipulations* (2011), *A Grammar of Luwo* (2014) and several other volumes. A book on language and emotion edited by her was recently published (*Consensus and Dissent*, 2017). She is a co-editor of the journal *The Mouth*.

Foreword

Tom Selwyn

Hazel Andrews frames this volume within a tradition of anthropological work on liminality – material and symbolic spaces of transition between leaving one social or geographical status and joining another. Authors associated with this genre include Arnold Van Gennep (2010), Victor Turner (1967), Reza Masoudi (2018), Hazel Andrews and Les Roberts (eds) (2012) and Andrews (2011) herself. This Foreword builds on this work while setting its compass towards the specific relationship between liminality, travel and Brexit. Along the way and taking the chapters of the present volume as starting points, it asks what contribution the anthropology of travel, tourism and pilgrimage (hereafter ATTP) is making to Brexit studies.

We may thus begin with a question (to be addressed as a single one) that has three interrelated parts. What sort of 'liminal spaces' has Brexit brought us to? How do these differ from anthropologically more familiar liminal spaces? What consequences follow from the differences?

Liminal spaces routinely mark passages between one identity and another (unmarried to married, for example) often using territorial metaphors to do so (as is the case with liminalities traversed by tourists as they leave home, spend time away on holiday and return). Almost invariably the identities engaged in rites of passage are known, recognised and celebrated by those involved. In the case of Brexit, however, the points of departure and arrival are very far from being known or recognised. As Catherine Palmer implies in Chapter 3, it is unclear what the nature of 'British' identity post-Brexit will look like. We could add that the nature of such identity pre-Brexit is also unclear. Furthermore, as Masoudi states in Chapter 13, Britain is walking into an unknown geography facing an equally unknown global status in the world. As he adds, the 'configuration of the kingdom (will be) increasingly fragile and fragmented', one implication being that 'Britain' itself may cease to exist – at least in its present 'united' form.

Masoudi's notion of fragmentation is pursued in Vivian Gornik's enquiry in Chapter 4 about whether the concept of British 'heritage' can reflect, contain and truthfully express the country's multicultural reality. It is a question that overlaps with Mark Casey's analysis in Chapter 7 of the sociocultural consequences of two television shows set in the Spanish resort of Benidorm. He argues that the 'racism, classism, dislike for Spanish culture and belief in the superiority of "the British" echoes themes to be found in the campaign to leave the European Union'. In the latter case, the cultural dispositions Casey observes in Benidorm find varieties of expression in political and media utterances within the culturally diverse topography of Britain itself. Jennifer Frost and Warwick Frost's depiction in Chapter 5 of royalty and its associated rituals such as the 2018 royal wedding of Prince Harry to Meghan Markle 'rooted as these are in a nostalgic imagery of Britishness, with the Royal Family at the heart of that narrative' is pertinent. Their work leads seamlessly into a question about whether the widespread sense of the importance and enduring qualities of traditions symbolically celebrated in royal rituals are enough to achieve what various politicians have rhetorically sought in recent electoral jousts, namely the bringing together of a culturally diverse country. There is a distant echo here of comparable questions asked by Bernard Cohn (1981) in his discussions of the role of royal rituals in Imperial India. He argued that such rituals made valuable symbolic contributions to the legitimation (until 1947 and Indian independence) of the ruling power of the British Empire by symbolically placing traditional Indian rulers within a ritual order in which the British king was paramount. The question Frost and Frost pose is whether the royalty of Brexit Britain can pull off the same feat and, if so, for how long.

Returning to our initial question, then, our authors suggest that the liminal spaces of pre- and post-Brexit Britain are full of unclarities in a number of identifiable respects. These include questions about our existing and future identities, our relations with others (including the question of which others), the post-Brexit name of the country (will there be a 'Britain' after Scottish independence and Irish unification?), the role of royalty in ensuring unity out of diversity and so on. Central to all of these issues is the nature of the nation itself. Is it a cosmopolitan space in which many cultural traditions flourish side by side or is it a version of the kind of narrow, white, perceived 'English' landscape with memories of empire and mythical expectations of recovered post-Brexit 'greatness' promoted by media outlets such as the *Daily Express*, the *Daily Mail* and the *Daily Telegraph* newspapers? Casey's chapter, like Andrews's (2011) ethnography, both describe nationalist expressions in, or about, Benidorm and Magaluf that suggest that the playgrounds of mass tourism are rich fields for a British nationalism of the latter kind: and we know from Victor Turner (1982) about the 'human seriousness of play'.

The first journey to be made, therefore, is the route through these at once playful, serious and unknown liminal spaces to a new, known, territory. At such a moment, people look to a leader to guide them.

Alexandra Pimor's reflections in Chapter 2 about Robert Schuman's notion of the 'spiritual unity' of European peoples through their membership of the European Union (EU) raise questions about the peaceful co-existence and interculturalism that a 'European consciousness' brings to a continent historically scarred by war and now a focal point of global movements of refugees and others seeking new homes, lives and work in its midst. Schuman foresaw a Europe in which people took pleasure in travelling, studying, making friends and building new networks throughout the continent. His Europe was one of cultural, philosophical and artistic unity and cooperation. This was to be the core of a European consciousness born out of the free movement of persons, ideas and knowledge. Such a vision contrasts with the historically 'semi-detached' nature of the British relationship with Europe. Successive Conservative governments have held a long-term ambivalence towards the European project (the Thatcherite promotion of the single market followed soon afterwards by rejection of the European social model, for example) and it is this ambivalence that eventually led to Brexit and the near total detachment from the EU that Brexit seems rhetorically to promise. Victoria Hooton pursues one of the consequences of this disengagement, namely the practicalities of the 'associate EU citizenship' that UK citizens may be able to claim post-Brexit. This proposed legal mechanism enables UK citizens to opt-in to EU citizenship and a legal framework that allows for free movement. Hooton warns that it will involve the effective downgrading of existing legal rights held by all full EU citizens. While it may make it easier for some 'educated, skilled or wealthy' citizens to exercise their freedom, those with fewer such attributes will find it more difficult. Such an outcome will amount to UK associate citizens being 'divorced from the values attached to EU citizenship's symbolic status'.

Given that 'Take back control' (of 'borders, money and laws') was, and remains, a leading slogan of the Brexiteers, and that one of the implications of this item of Brexit-speak is that leaving the EU promises all manner of 'freedoms' from the supposed regulations of 'Brussels', it is no surprise to find that at least five of our chapters are concerned with freedom and freedoms in varieties of different but complementary contexts. In Chapter 8, Andrews reflects on 'freedoms' enjoyed by holidaymakers in Magaluf, ground down, as some of them explain, by the lack of freedom in their daily working lives at home. But we would wish to observe that 'holiday freedom' is hardly ever an unfettered one. Andrews (2011) herself has skilfully shown how Magulufers are drenched with nationalist propaganda and incorporation into consumerism. Stan Cohen and Laurie Taylor (1976) included tourism as one of the 'escape routes' out of such a commoditised world but concluded that the liminality experienced by some types of tourist practices does not deliver escape from the working routines of neoliberal economic regimes but actually reinforces them. One of the underlying concerns of Sabina Owsianowska and Magdalena Banaszkiewcz's Chapter 6 is the degree of freedom, or lack of it, encountered by migrants to the UK from Poland and other central European countries. As Sudarshan (2017) has shown, following the increased incidence of hate crimes in the UK post-referendum, the 'link between Euroscepticism and racial or bigoted animus... towards all marginalized groups, including those from Poland, some of which were not even related to Brexit complaints, spiked... Brexit was a mandate for hateful behaviour'. Marcus Stephenson and Shaun Goldfinch's treatment of the 're-imagining of the Commonwealth' in Chapter 10 points out the flawed ambiguity of, on the one hand, rhetorically brandishing the desirability of the Commonwealth (all of it) being our future trading partner while, on the other hand, steering us towards some sort of superior Anglophone trading network. As in Hooton's chapter, there is a *trompe-l'oeil* here that fits well into Andrews' discussion in her Introduction of Brexit and magic. The intriguing discussion by Maria Amoamo in Chapter 11 about the actual outcome of the apparent invitation being made by the Brexit process to the UK overseas territories – in the main small islands such as Pitcairn Island, her case study – to sit at the world's top trading table follows Stephenson and Goldfinch's concerns with the Commonwealth. Will Brexit *really* free the island territories from their subordinate position in the world's economy or will they *actually* remain as economically vulnerable as they are today? Finally, Anne Storch and Nico Nassenstein's investigation in Chapter 9 of African souvenir sellers offers us a timely reminder that an under-researched, often neglected and taken-for-granted category of tourism worker can reveal much about how the ripples of Brexit extend far away from the hot houses of Westminster and Brussels, in this case into Africa as these 'polyglot players perceive and estimate the consequences of having to cope with potentially decreasing numbers of British tourists in a semiotically complex and competitive space'. Anthropological interest in this chapter includes a methodology based 'on less conventional sociolinguistic and anthropological methodology such as walking and "small stories", collected among communities of largely undocumented migrants'. They assert that their study 'takes those who are usually considered to be at the periphery of the tourism economy... seriously as very central protagonists'. We could add two observations here. The first is that analyses of tourism-related economies need to pay as much attention to the 'informal' economy as to the 'formal' one (Çakmak *et al.*, 2018; Crick, 1994). The second is that, for the sake of ethnographic accuracy, workers in tourist-related informal economies whose remittances from diasporas are often central to the economies of the Global South need invitations into the mainstream of ATTP.

How, then, might we summarise what our authors tell us about the relationship between Brexit and ATTP in such a way that allows us to identify the principal themes of their essays? Taking a main theme from each of the 12 essays considered here (i.e. Chapters 2–13), we may locate 12 themes. Pimor (Chapter 2) speaks of *displacement* referring, in particular, to the withdrawal of the UK from the project of European unity and suggesting other forms of displacement too. Palmer (Chapter 3) talks of post-Brexit *fragmented identities* in the UK while Gornik (Chapter 4) draws our attention to the problem of subsuming British *cultural diversity* into a project that seems (by some at least) to be built on ideas of cultural homogeneity. Frost and Frost (Chapter 5) raise the question of the possible role of royalty playing a part in the processes at work in searching for *a leader/guide* in an uncertain cultural landscape. Owsianowska and Banaszkiewcz (Chapter 6) reflect on post-Brexit *freedom of movement*, particularly when related to migrants from Central Europe to the UK. Casey (Chapter 7) talks of Brexit and the *subordination of the Other* (his 'Benidorm complex'), Andrews (Chapter 8) of the elusive *freedoms* of the holiday, while Storch and Nassenstein (Chapter 9) open up the issue of *Brexit in the world* (in their case through the medium of African souvenir sellers in Mallorca) and Stephenson and Goldfinch (Chapter 10) and Amoamo (Chapter 11) of what could be termed the *Commonwealth trompe l'oeil* and *overseas territories three card trick*, more prosaically expressed as the possibility that some invitees to the top trading table may find themselves left behind in the banquet's ante-room as the reality emerges of a white Anglophone conglomerate of trading nations, led by the United States, skimming off the trading cream for itself. Hooton (Chapter 12) raises the prospect of an emergent post-Brexit *two-tier European citizenship*. Finally, Masoudi (Chapter 13) stresses the *unknown landscape* into which the Brexit process is taking us.

Returning to the question with which we started out, what consequences might we draw from the anthropological and ethnographic work presented here, both for the relationship between Brexit and ATTP and for Brexit studies more generally?

Several chapters refer in a variety of ways to a characteristic of Brexit practices with which we have become familiar since the despatch of Theresa May's 'hostile environment' vans into areas of the UK thought by the government to be homes for 'illegal immigrants'. In the cases of Hooton's concerns about the possible emergence of a two-tier European citizenship (one for the rich, another for the poor) or Casey's 'Benidorm complex', to take just two examples, our writers inform us of the links between Brexit processes and the rhetorical and actual subordination of the Other. The British Home Office, presumably ordered by the then Home Secretary, named the 'hostile environment' project 'Vaken' which the *Guardian*'s Simon Hattenstone (2018) explained derived from the German rallying cry 'Deutschland Erwake', 'Germany Awake', a slogan emblazoned on banners flown at the Nuremberg rallies in the 1920s and 1930s. May's own Vaken image was of handcuffs with three textual messages. 'In the UK illegally?' asked one, '106 arrests last week' announced a second, 'Go home or face arrest' said a third. The Vaken project was part of the rhetorical stream that led to the 'Windrush Scandal' and, given Nigel

Farage's 'Breaking Point' image at his launch of his Brexit activities on the day of the murder of Jo Cox MP, Brexit itself.

One noticeable feature of Brexit rhetoric as heard in the UK that the present volume brings into the open is the more or less total insularity of Brexit-speak. Despite the fact that we are promised a 'global trading future' after our emergence from the EU, we actually hear very little about how the economies and societies in other parts of the world will be affected by Brexit. In our collection, however, Commonwealth countries, workers from an African diaspora, UK overseas territories (who mentioned Pitcairn Island in their Brexit speeches in the UK before the referendum?), the countries of Central Europe and Mediterranean tourism-related resorts, all occupy central places. This may not be surprising given the anthropological nature of the collection, but it is worth stressing that although the nature of the general Brexit discussion has been dominated by references to 'migrants', especially 'illegal immigrants', it has contained little about economies elsewhere in the world. We are, of course, told that we will 'control our borders' by selecting only those people as immigrants who will serve *our* economy. The contempt for the rest of the world that this lacuna alone conveys is matched and confirmed by such contemptible actions by Parliament as the subtraction of the rights of children to join their parents in the UK from the EU Withdrawal Bill.

This Foreword started with a consideration of the relation between liminal spaces and Brexit. At this point, we may follow our authors' lead by identifying a quartet of terms that effectively summarise the version of 'Brexit Britain' that underpins their studies: displacement, detachment, fragmentation and fragility. As all our authors state, the UK itself is being displaced: from the European project of further integration as well as (in a variety of ways some of which have not yet become apparent) from continental Europe itself. At the same time, Brexit has brought with it a geography of fragmented identities and increasingly fragile bonds between constituent parts of the country. Brexit has guided the people of the UK, and the UK itself, into liminal spaces in which detachment and separation are dominant motifs. The fragmented spaces into which we have been drawn by the extreme right architects of Brexit are now governed by widespread senses of these overlapping terms. A man recently made redundant from the closing of the Goodyear factory in Wolverhampton, recorded by Jim O'Neil (2017) for his 'Fixing Globalisation' TV series, explains why the town voted to leave the EU. 'All I have known', he said, 'in 38 years of working in this country – all my working life – is short-time working, recessions, people losing their jobs, factory after factory closing... This region has been stripped down and taken abroad'. He described the food banks, the homeless, the 'people queuing up 60-deep outside Citizens Advice worrying about bailiffs' and concluded 'This is my town, and we've got to do something about it – and that's why people voted Brexit as a protest vote'.

The origins of Brexit go back a long way, at least to the time of the Conservative project under Margaret Thatcher which involved the shrinking of the state, reduction of taxes, deregulation, privatisation and the reduction of public services (including public housing), loss of heavy industry, demolition of union strength and influence, promotion of financial services, austerity, hostility to migrants, the privileging of national identity above others and the seeking of closer relations with the United States. Brexit must have appeared as the prize of all prizes: all these aims could be achieved by one single blow. And the 'people' (as in 'Will of the People') could imagine that the project was none of the above but rather the fulfilment of the seductive myth of 'taking back control'. The question then arises about the conditions in our contemporary world that give rise to the generation of, and belief in, such a myth. Hannah Arendt's (1951) *Origins of Totalitarianism* has an answer. If society and social structures are broken (like those as bruised and fragmented by austerity as those in the UK) and citizens have become more or less isolated individuals, then the ground is ready for the rise of powerful myths that serve to give us what we lack in practice: a sense of purpose and agency even if this is known to be imaginary. As Arendt says, totalitarian regimes succeed because isolated, atomised individuals get a 'sense of having a place in the world'.

What will future researchers in the field covered by this volume do next? We may find some answers by emphasising once again what our authors have taught us about 'Brexit Britain' and then using Thatcher's famous myth about there being no such thing as society to trace the way ahead.

As noted in some detail throughout this Foreword, our authors have painted a portrait of contemporary British society and culture that is framed by senses of fragmentation, fragility, displacement and detachment. They have recorded how these senses are manifest by describing how the country is being detached from its neighbours, how media rhetoric constructs migrants as unwelcome others, how feelings about the past seem as uncertain as those about the future: in short how our collective identity is fracturing as we become disconnected individuals in an unknown and seemingly unknowable cultural, economic, political and social landscape.

None of this should come as a surprise since this has been the plan for over four decades. The contemporary heirs of the Thatcher project, presently in power and looking to remain in power for years to come, are leading Britain out of the EU in order to complete the political revolution the lady promised since her emergence in 1979. The question for us and our colleague researchers is thus familiar and precise. What is our response to Thatcher's assertions in her famous *Women's Own* interview in October 1981? In this interview she said: 'we've been through a period where too many people have been given to understand that if they have a problem, it's the government's job to cope with it. "I have a problem, I'll get a grant". "I'm homeless, the government must house me". They're casting their problem on society. And, you know, there is no such thing as society. There are individual men and women, and there are families... no government can do anything except through people, and people must look to themselves first. It's our duty to look after ourselves and then, also to look after our neighbour. People have got the entitlements too much in mind, without the obligations. There's no such thing as entitlement unless someone has first met an obligation'.

As researchers into the relation between ATTP and Brexit, our own obligation is to look for the recovery of that sense of the social we have learnt from travel itself. For present purposes, we may take our lead from four sources: the authors of this volume; aspects of the long history of travel and travel writing some of which is summarised in Selwyn (2018); and two outstanding commentaries that seek to respond to Thatcher's social and cultural nihilism. These are Guy Standing's (2019) *Plunder of the Commons* and the volume edited by Ed Mayo and Henrietta Moore (2002) *Building the Mutual State*.

Connecting up and summarily piecing together some of the foundational insights of the above sources gives us a working basis from which to launch a vision for further research. Schematically put, if our authors have shown a Brexit process that is leading us into states of fragmentation (truth into lies, knowledge into fake news, hospitality into hostility, open into closed borders, confident identities into fragile ones and so on), much of the travel writing we know (from Heroditus to Montaigne and Francis Bacon to Malinowski himself) speaks of travel as the harbinger of ethnographically and empirically based knowledge, as being the means of discovering forms of journeying that bind the individual traveller into the world, as a way to experience the social and the pleasures of hospitality and so on. All of this provides an intellectual, political and practical background to identify six areas of future research in an ATTP field under the shadow of Brexit.

Firstly, there is the work of professionals – amongst them doctors, councillors, lawyers, academics and journalists – who work at the cutting edge of the 'hostile environment', bad parent of Brexit, by working in such contexts as detention centres, courts and media outlets to uphold principles of hospitality, of possibilities of regional and global identity, of open borders, of the socially creative liminal spaces that travel can open up. Secondly, there are the growing number of local municipal schemes (well away from Westminster) to provide hospitality to migrants (Selwyn, 2019), in the process countering the hostility of the Brexiteer press and its often anonymous social media followers. Here, the role of government in providing help and support to such municipal initiatives comes into sight, as Mayo and Moore (2002) carefully explain in their work on the mutual state. Thirdly, there is the work of architects, urban planners, guardians of public spaces and others, who work to uphold principles inherent in the ideas and values of the 'Commons' (spaces that, as Standing [2019: 27] tells us are 'general, universal, shared, held in common'). Associations successfully fighting library closures are a case in point. Fourthly, there are those parts of the leisure studies field held together by theories and practices that celebrate the contribution many kinds of leisure activities (from choirs to dance venues, faith-based associations to public parks and gardens) make to the strength and well-being of community and society. In our own ATTP field, for example, the growing amount of travel writing (from Jean Jacques Rousseau's, 2000 [1782] *Reveries of a Solitary Walker* onwards) and research on the sociocultural implications of walking tours is a case in point. Fifthly, there are a very large number of civil associations preoccupied with travel and social thought. Lobby groups such as 'Equality in Tourism', associations such as the company offering tours of 'Unseen London', led by homeless guides are examples. Sixthly, there is an expanding amount of university and literary work in the field of ATTP that regards travel as a possible crucible for social and cultural creativity. University courses and summer schools (including those at SOAS) are all examples. Following Launay's (2010) brilliant history of early travel writing and the impact it has had on the development of social science, such academic and literary work reminds us that the experience of travel and the testimony of travel writers have made foundational contributions to the building of anthropology itself.

We may conclude by recognising that the achievement of this book is twofold. First of all, its authors have guided us into some of the darker liminal spaces of the Brexit process, haunted as these are with senses of the unhomely and fragmented – Masoudi's 'unknown landscapes' or Standing's (2019) 'plundered commons' – from which senses of society and the social have been banished in order for the emptying spaces to be filled with individuals increasingly detached from each other. But, in their very identification of this dark world – one dominated by 'deals' rather than exchange – our authors have also (if we can put it this way) implicitly challenged readers and researchers to look into those features of travel (hospitality, curiosity for knowledge of the world, discoveries of modes of commonality and so forth) that can reveal possibilities for those same liminal spaces becoming sites from which the commons can be recovered.

References

Andrews, H. (2011) *The British on Holiday. Nation, Identity & Consumption*. Bristol: Channel View Publications.
Andrews, H. and Roberts, L. (eds) (2012) *Liminal Landscapes: Travel, Experience and Spaces In-Between*. London: Routledge.
Arendt, H. (1951) *The Origins of Totalitarianism*. New York: Harcourt Brace.

Çakmak, E., Lie, R. and Selwyn, T. (2018) Informal tourism entrepreneurs' capital usage and conversion. *Current Issues in Tourism* 22 (18), 2250–2265.

Cohen, S. and Taylor, L. (1976) *Escape Attempts: The Theory and Practice of Resistance to Everyday Life*. London: Allen Lane.

Cohn, B. (1981) *Representing Authority in Victorian India*. Princeton, NJ: Princeton University Press.

Crick, M. (1994) *Resplendent Sites, Discordant Voices: Sri Lankans and International Tourism*. Chur, Switzerland: Harwood Academic.

Hattenstone, S. (2018) Why was the scheme behind May's 'Go Home' vans called Operation Vaken? London. *The Guardian*.

Launay, R. (ed.) (2010) *Foundations of Anthropological Theory*. Chichester: Wiley.

Masoudi, R. (2018) *The Rite of Urban Passage*. Oxford: Berghahn.

Mayo, E. and Moore, H. (eds) (2002) *Building the Mutual State: Findings from the Virtual Think Tank*. London: New Economics Foundation.

O'Neil, J. (2017) Fixing Globalisation. BBC, London.

Rousseau, J.-J. (2000 [1782]) Reveries of the solitary walker. In C. Kelly and A. Bloom (eds and trans.) *The Collected Writings of Rousseau* (pp. 8–93). Lebanon, NH: Dartmouth College Press.

Selwyn, T. (2018) Tourism, travel, and pilgrimage. In H. Callan (ed) *The International Encyclopedia of Anthropology* (pp. 125–156). London: Wiley.

Selwyn, T. (2019) Brexit, Grenfell, Windrush, and the mooring, un-mooring, and re-mooring of home. In S. HadžiMuhamedović and M. Grujić (eds) *Post-Home: Dwelling on Loss, Belonging and Movement. Ethnoscripts*. Hamburg: University Press.

Standing, G. (2019) *Plunder of the Commons: A Manifesto for Sharing Public Wealth*. London: Penguin Random House.

Sudarshan, R. (2017) Understanding the Brexit vote: The impact of Polish immigrants on Euroscepticism. *Humanity in Action*, New York.

Turner, V. (1967) Betwixt and between: The liminal period in rites de passage. In V. Turner (ed.) *Forest of Symbols: Aspects of the Ndembu Ritual* (pp. 23–59). Ithaca, NY: Cornell University Press.

Turner, V. (1982) *From Ritual to Theatre: The Human Seriousness of Play*. New York: PAJ Publications.

Van Gennep, A. (2010 [1960]) *The Rites of Passage*. London: Routledge.

1 The Magic and Liminality of Tourism and Brexit

Hazel Andrews

Introduction

> An emancipated society is one that achieves coexistence in difference.
>
> Jackson, 2013: 9

Preamble

When I was invited to either write a monograph or compile an edited collection on the theme of tourism and Brexit I could not at first imagine what such a book would look like, given that Brexit (that is the withdrawal of the UK from the European Union [EU]) had not – and, at the time of writing has not – happened. How could the relationship between the activities that make tourism such an important sector of the global economy and the national economies of many countries and something that has yet to happen be understood? This edited collection is an exploration of the relationship by picking up on some of the key themes that are relevant to both the practice of tourism and the Brexit debate; that is travel, borders and identity.

Both tourism and Brexit invite reflection on travel whether that be in the actual travel of tourism, the association of the EU with travel through the freedom of movement and the associated perceived problem of immigration, or the metaphorical travel of the Brexit process moving the UK out of the EU; from one state to another. Tourism and Brexit both concern borders, crossing them on the one hand or re-erecting them on the other, and both are deeply concerned with questions of identity. Additionally, both tourism and Brexit belong to the social world, having arisen through it and, therefore, reflective of it.

Brexit whys and wherefores

This introduction and this book do not seek to explain why Vote Leave won the 2016 referendum. Others have started to do that and are too numerous to consider here.[1] However, it is worth outlining some examples. For instance, archaeologist Andrew Gardner (2017: 3) argues that 'Brexit can be seen as the culmination of the collapse of the British empire, and transformation of British identity, in the post-Second World War era', and with heritage studies expert Rodney Harrison discusses the implications of Brexit for archaeology and heritage, noting the rise in 'new nationalisms' (Gardner & Harrison, 2017). As they point out, this is not the preserve of the UK, but a global phenomenon. Hugh Gusterson (2017), for example, links Donald Trump's election as US president with Brexit in what he describes as 'the upsurge in nationalist populism'.

Questions of national identity certainly played a role in encouraging the vote to leave the EU, as some sentiments expressed in its aftermath indicate. In the UK, there was a rise in racist abuse post-referendum (for examples, see Stein, 2016) which, according to Madeleine Reeves, can be laid at the door of Vote Leave. As she argues, their '"call to take back control" has given form and solidity to undercurrents of fear, disillusion and xenophobia' (Reeves, 2016: 480). Although Reeves may have a point that Vote Leave capitalised and gave voice to latent sentiments concerning identity insecurities, part of the problem lies much deeper than that within neoliberal politics and the accelerated capitalism of the principles of the free market unleashed by the Thatcherite Conservative government of the 1980s and continued in the New Labour Project led by Tony Blair in the 1990s. As Gillian Evans (2017: 216) points out, there was 'the complete failure of the New Labour government (1997–2010) to address and represent the interests of its traditional supporters'. Tom Selwyn illustrates the point in hand in relation to tourism and heritage. In a study that examines the relationship between self and other that is manifest in tourism through the many images that are used to attract tourists, Selwyn examines the post-industrial landscape of a UK once based on manufacturing and production industries such as coal mining, rather than a service economy. With specific reference to a former mine in Wales, now a heritage attraction, he asks, 'Do *they*, the beaten but here heroic miners, belong to the same lineage as us? If they do... how did we keep faith with them when the combined forces of technological and economic change and a vindictive government were arraigned against them? Some might say that we abandoned them until it was safe to build them a World Heritage Site' (Selwyn, 2010: 208). Certainly a valid point and one to keep considering as the post-referendum analysis rumbles on. However, we need to remember that those who voted leave were not just composed of a disaffected or left-behind working class. And, it should also be remembered that not all people identifying as working class or with working-class roots voted leave (see, for example, Knight, 2017).

We're all Europeans now

Of course, the issue that fuelled the Brexit debate was related to the UK's uneasy relationship with the EU[2] which was argued by Vote Leave to be subsuming UK identity and sovereignty in place of the development of a European identity. Anthony Pagden (2002: 33) points out that, 'The identity of "Europe" has always been uncertain and imprecise', that 'like all identities it is a construction, an elaborate palimpsest of stories, images, resonances, collective memories, invented and carefully nurtured traditions'.

In her discussion of the Italian city of Trieste, Luiza Bialasiewicz argues that the forging of European identity-making began at the end of the Second World War. She notes that as a city on the border with a communist country (the former Yugoslavia), like Berlin, which also directly adjoined a communist state, Western Europe rubbed directly with its non-Western others and that '"Europeanness" had to be daily re-inscribed' (Bialasiewicz, 2009: 320). However, the idea of Europe has deeper roots than that of the mid-20th century. In one of his Reith Lectures for BBC Radio 4 in 2016, philosopher Kwame Anthony Appiah points to the creation of the notion of Europeans, and by corollary 'Western' civilisation, as it rubbed up against another 'other' in the form of Islam (for example the Turkish siege of Vienna in 1683 and the Moorish rule in Spain from 711 to 1492). In this creation of opposition, religious figures, chroniclers and philosophers have shaped the idea of the European based on ancient Greece: 'Hegel, the great German philosopher, told the students of the high school he ran in Nuremberg, that foundations of higher study must be and remain Greek literature in the first place, Roman in the second' (Appiah, 2016).

The invention of traditions that Pagden makes reference to, and the invention of a shared cultural lineage in the form of Greek and Roman ancestry to which Appiah alludes, speaks of the creation of a narrative of Europeanness also identified by Cris Shore (1993: 779) in his examination of the processes of European integration in which he highlighted 'the European Commission's attempts to forge a supranational "European Identity" through the development of symbolic measures'. Many will be familiar with such pan-European ventures as the European Song Contest, the European Capital of Culture, the adoption of Beethoven's 9th Symphony Ode to Joy as the European anthem and, for scholars, research funding that seeks to build academic partnerships in consortiums working on pan-European issues. Some of these initiatives pre-date the European Commission's attempts since 1977 to develop 'a "cultural policy" which has helped to boost people's awareness of a European cultural identity' (Shore, 1993: 779). In terms of the idea of community building and developing common goals, such initiatives have their merits; however, as Shore (1993) argues, a

> problem with the Commission's attempts at defining a cultural policy lies in reconciling 'spontaneity' with 'instrumentality'. However well intended, Article 128 of the Maastricht Treaty… nevertheless raises fears

of excessive bureaucratic intervention. Institutional attempts to 'foster', 'encourage' or 'promote' the diversity of European cultures are bound to be seen as creeping centralization. (Shore, 1993: 794)

If the warning signs were there in 1993, it is easy to imagine how in the 23 years that elapsed until the UK referendum in 2016, the fears of losing national identity could become both more expressed and more entrenched.

Tourism and Brexit

Prior to the referendum, tourism industry lobby groups had forecast some drawbacks to being outside the EU, including, for example, a reduction in business travel and the ending of EU funding in support of tourism development (Tourism Alliance, 2016). However, tourism is not merely composed of business interests and practices. There are business functions that facilitate the practice of tourism; however, to leave argumentation solely in the hands of business interests would not only miss the point of tourism as a sociocultural phenomenon and all that goes with that (for example, ideas of cultural exchange or even peace building rightly or wrongly attributed to the merits of international tourism – see the World Tourism Organisation [WTO] Manila Declaration, 1980), but would also abandon tourism to the vagaries of advanced capitalism concerned as it is with audits, profits and market supremacy without considering the very humanness of the desire to travel, socialise within, experience, see and embody the wider world. It would deny the very livingness of being alive, for travel-tourism is the practice of life and life the practice of travel. We travel through life and life takes us forwards, or backwards, through our imaginations, hopes, desires and memories. Thus, the intermingling of tourism and Brexit is not just about balance sheets, employment and numbers of people travelling, but it is also about expressions of how we see, feel and inhabit the world, how the world sees and feels us.

It is noteworthy that one of the most immediately identified issues for tourism in the aftermath of the result of the referendum is that the UK is seen as a less welcoming, a less hospitable country to choose to visit than before June 2016 (Andrews, 2019). It is reasonable to suggest that any losses in business terms from inbound tourism could be balanced with the idea that people will choose to holiday at home thereby boosting local economies. Contrary to the idea that the UK will attract fewer visitors, the *Daily Express* tabloid newspaper ran the headline 'Britain's £26BN Brexit Tourism Boom' on 3 January 2020 with a claim that 2020 would 'break records' (p. 4) and '[the] treasured isles [would] lure 39m guests' (p. 5). However, this forecasting needs to be tempered with the understanding that how hospitable a place or peoples are perceived to be is of great importance to the contemporary manifestation of hospitality as a commercial enterprise (Lashley & Morrison, 2000) and an important 'tool' in customer relations in tourism (see Andrews, 2000). Further, economic business arguments are not enough for the sociality that is the bedrock of reciprocal forms of exchange – of which hospitality is one manifestation – and the very basis of society (Mauss, 1954).

Brexit does not simply mean leaving the EU, but rather it is a pregnant signifier that carries so many ideas, so many ideals. For many, it is about a journey to a better place outside of the apparent control of perceived faceless bureaucrats who dwell in an imagined land of faraway. This land of faraway is not just about geographical location, but relates in the UK to questions of identity, values and habitus, all of which are seen as under threat in the form of the EU other and its perceived incursions (maybe read invasions) across the real and imagined borders between 'us' and 'them' to tell 'us what to do'. For others, the creation and membership of the EU represent freedom, cultural exchange, economic security and a sense of community. Freedom here mainly relates to the freedom of movement; the ability to travel easily across the borders of the different countries that comprise the EU, whether to holiday, visit friends and family, engage in business, take up a new job or retire to somewhere warmer and with a perceptually more attractive lifestyle than that offered in the quotidian world of one's country of birth.

Tourism, in general, is premised on offers of a better life elsewhere in which freedom can be found and desires satisfied. In its manifestation as physical travel, it requires the crossing of boundaries even if that is just across the threshold of one's own home, and it offers a period (or multiple periods) of transition in which the self is betwixt and between, in a liminal state. In this respect, both Brexit and tourism have something in common, but they pull in different directions in relation to ideas of travel. Tourism premised on the obvious need to move, cross borders and Brexit with its promise to end freedom of movement within the EU, that constitutes a disincentive to travel.

Liminality, myths and magic of tourism

That tourism is linked to myth, magic and liminality is well-rehearsed in the academic tourism literature. As noted, touristic practice offers opportunities for transition and can be likened to the ritual process in terms of leaving somewhere (marking a period and space of separation) and then on return being reincorporated into the home world having undergone some kind of transformation, whether that be through the acquisition of a suntan or engagement with different kinds of experiences from that found in the everyday (Van Gennep, 1909). Nelson Graburn (1976) likened tourism to a ritual process in the emerging literature on the anthropology of tourism in his discussions of tourism as a sacred journey;

and Edward Bruner (1991) examined tourism as a means for transformation of self. In addition, the link between tourism and liminality with the latter's allowance for transgression, has infused the work of scholars such as Rob Shields (1991). In relation to the idea of liminality as part of a process of becoming, it has laid foundations in my work (Andrews, 2009, 2017) to discuss the latent possibilities of being that tourism practices can facilitate.[3] Elements that make up the tourists' journey have also been discussed as liminal, for example Huang *et al.* (2018) on airports and Pritchard and Morgan (2006) on hotels.

The magical qualities of tourism lie, in part, in its ability to enchant through the stories it tells and the myths it creates. Through the identification of the spaces of touristic practice and the attendant promotion of such places through the various forms of images that emerge in tourism literature – brochures, postcards and so on – and other forms of media, Selwyn (1996) has identified the mythical nature of tourism. He points out that his discussion of myths is derived from a Lévi-Straussian tradition. By this, he means that the stories woven around places and peoples of interest to tourists 'seem simultaneously to reveal and conceal' (Selwyn, 1996: 3). That is, on the one hand, places such as Nepal are shaped for tourists' imaginations based on their 'preoccupations, motivations and understandings about… harmonious social relations, ideas about community, notions of the whole' (Selwyn, 1996: 3) but which at the same time obfuscate the actual living conditions of the people who inhabit the destinations (Hutt, 1996). As such, the success of tourism is, in part, based on its ability to enchant (Selwyn, 2007) the potential tourist, to weave a spell that makes life look more attractive elsewhere. David Picard (2011) explores the development of the tropical island of La Réunion through different concepts of magic. The island, he argues, becomes like a magical pleasure garden in which ideas of the destination and its people are cultivated as if they were plants in a garden. In earlier writing, Picard (2010: 150) argues that tropical islands have a specific role in the modern world system and that in their promotion as tourist destinations in travel marketing material, they 'embody some essential dreams, desires and enchantments embedded in the Western imaginaries that can be invoked through tourism'. Andrews (2014) has referred to the ability of the narratives of tourism destinations to enchant tourists so that the symbolic violence that underpins the portrayal of normative gendered relationships in two Spanish resorts is masked by fairy tale-like and swashbuckling heroic adventures of pirates occupying a liminal state.

The Magic and Liminality of Brexit

Magic

Following the outcome of the referendum in June 2016, the then Conservative Prime Minister David Cameron resigned, to be replaced by Theresa May whose mantra became 'Brexit means Brexit', a phrase that was repeatedly uttered by her during her disastrous[4] general election campaign of 2017. It was as if she were reciting an incantation that in its saying would invoke the UK's withdrawal from the EU; but unlike Searle's (1998: 115, in Moeran, 2016: 2) observation that 'saying something makes it true', this was not the case and May could not deliver Brexit. And nor did the saying 'Brexit means Brexit' give further enlightenment to what the actual consequences of the UK's departure from the EU would mean in practice.

Writing a few months after the referendum in 2016, Brian Moeran observes that the Brexit debate in the lead up to the actual vote could be understood in anthropological terms as magic. He argues that the referendum itself was a rite 'in which politicians can dress their magical part' (Moeran, 2016: 2). Drawing on Malinowski's (1948) remarks that magic and uncertainty are linked, Moeran outlines how the discourses associated with both sides of the EU debate contributed to an air of uncertainty. Indeed, both sides had attempted to divine by forecasting a future for the UK outside of the EU, on the one hand, and remaining in the EU, on the other. The remain side was called 'Project Fear' by Leavers with its predictions of economic hardship and associated difficulties of life external to the EU. Remainers also accused Leavers of fear-mongering as those who championed Brexit ran a campaign based on anxieties relating to the malevolent forces of the EU and threats of more immigrants – most notably from Turkey – to the UK should the country remain an EU member. The language that is often associated with perceived high immigration numbers, for example 'swamped' as used by former Defence Secretary Michael Fallon in 2014,[5] suggests that what is at stake is the very body and soul of the UK as the country is increasingly turned into an unrecognisable form.

The language of myth and magic was evoked by the pro-EU side in referring to the promises of Vote Leave as mythical, symbolised by, for example, the use of the word 'unicorn' to speak to the fantasy world of the future the Leavers were said to be conjuring up. As Amy Davidson Sorkin (2019),[6] writing in the *New Yorker*, noted, 'Anti-Brexit protesters have taken to wearing unicorn costumes' and the Irish Taoiseach Leo Varadkar argued that Leavers' claims that the issue of the Irish border could be solved by implementing new technology was a practice of faith in 'magical solutions'. In another example, the daily newspaper *The Guardian* ran an article based on the news that the departure date from the EU had shifted from 29 March 2019 to 31 October 2019, the date of Halloween, a ritual that is supposed to ward off witches and other evil spirits. Columnist Andrew Martin argued that it was Vote Leave who had been spreading fear in the threat from politicians including Jacob Rees-Mogg, Boris Johnson and Iain Duncan Smith that failure to deliver Brexit would result in the eruption of politically far-right sentiments (which post-referendum had already seen an increase). Martin likened Rees-Mogg to both a corpse and an undertaker in relation to his style of dress, labelled Johnson 'just an all-purpose nightmare' and suggested that Duncan Smith 'is another ghoul, closely resembling Nosferatu, and favouring Phantom of the Opera hats'.[7]

Brexit then cannot just mean Brexit, other than Brexit means a period of reflection and change. It is a disruptive force – for good or ill, depending on perspective – that is destined to bring transformation, and which already has. As a project in national self-reflection based on such a narrow majority win in the referendum for Vote Leave, it has clearly identified schisms between the constituent countries of the UK. Indeed, Scotland voted overwhelmingly to remain in the EU, as did Northern Ireland. By contrast, Wales and England voted to leave, but within these countries there are pockets of strong remain centres. The divisions exposed by Brexit mean that the union upon which the UK rests is ever more under threat with calls from the increased Scottish National Party majority to have another referendum on Scottish independence, and the possibility of a united island of Ireland now looked upon more favourably in its geographical North. Brexit, then, causes us to reflect on boundaries – restating borders between the UK and the EU, a perhaps less fluid border between Scotland and England should the former gain independence, and the potential for the total dissolution of a border between the Republic of Ireland and Northern Ireland. Such changes impact on how people understand their place in the world and as such their sense of self and identity.

Elsewhere, I have questioned whether this representation of division in the UK should come as a surprise given the expressions of differences, dislike and distrust I identified between the regions and countries in the UK at the end of the 1990s in my fieldwork to Mallorca (Andrews, 2017). What Brexit may have done is draw attention to these differences on a much wider national scale. Indeed, discussions following the June 2016 referendum have focused on the north–south divide, class, age and educational differences between those who have become known as Remainers and Leavers (see, for example, Brown, 2016 and Hobolt, 2016).

The uncertainty and political wrangling over Brexit have led to a change in the political landscape of the UK as the December 2019 parliamentary elections, which proved to be a *de facto* second referendum, have demonstrated. Despite years of austerity first initiated by a Conservative–Liberal Democrat coalition and continued by a Conservative minority government, and attempts by the opposition Labour Party to reposition the political debate to focus on the welfare state, it was, in part, the attraction of the simple slogans of '*Get Brexit done*' with '*an oven ready deal*' repeated mantra-like by Boris Johnson as part of his campaign rhetoric that turned the so-called Labour Red Wall of traditionally held seats across the English midlands and north to rubble, although in many of these places the newly elected Conservative MPs enjoyed only a very slim majority. In keeping with the idea of magic, it is noteworthy that in both the 2017 and 2019 election campaigns, Labour's ambitious spending plans were referred to as being funded by a 'magic money tree'.

As noted, the political rhetoric associated with Vote Leave prior to the June 2016 referendum played on many fears and insecurities related to ideas of national identity and control over destiny. Calls to 'Take Back Control', curtail immigration and 'Make Britain Great Again' all gained purchase, fuelled by a predominately right-wing media that has painted the EU as a threat rather than a builder of peace in Europe upon which some of the initial ideas for its creation were based. Indeed, writing in 1998 in the British tabloid *The Daily Mail*, columnist John Casey explained that the story of the Tamworth Two (two pigs that had escaped their fate at an abattoir in Malmesbury, west England – see Andrews [2011] for the full story) had fascinated the British because they represented something about the national character which distinguished 'us' from our European neighbours. Of course, what the national character is, if in fact there is one, is a matter for debate. As has frequently been discussed, British and English are often used interchangeably and there is a tendency to subsume other identities and ethnic descriptors into one that is not representative of a whole (Kumar, 2003).

Thus far, we can identify that the result of the 2016 referendum and the ensuing debates within the UK have already effected some changes, a country no longer divided on traditional political lines of left and right, changes in the way that the country is perceived, changes in demographics with many continental European citizens returning to their countries of origin, shifting national identities as many seek out an Irish ancestry that will give them the right to an Irish passport, in turn allowing them the continuation of freedom of movement currently under threat for British citizens from Brexit; and, by corollary the freedom of movement to the UK by EU members is also at risk. Overall, however, what Brexit will mean to the inhabitants of the UK can only be predicted. The arguments and uncertainty are like the fogs of a crystal ball that require someone with special powers to clear them and see the future. This leads us to the realm of magic. And, standing on the threshold of the monumental changes that Brexit may bring to the UK as it transitions from the state of being an EU member to another outside of the EU makes the process one of liminality.

Liminality

In some ways, the UK's current political state in relation to Brexit is both liminal and not liminal. For those familiar with the concept of liminality, its roots in Arnold Van Gennep's rite of passage and its development by Victor Turner during the 1960s and beyond, the link with notions of 'betwixt and between' (Turner, 1967) may seem easily applicable to the UK because, as I have already noted, the country is in a state of transition as it leaves the EU. An element of liminality relates to its processual nature and of possibilities; as Graham St John (2008: 4, emphasis in original) notes, 'The processual project recognizes that society is in-composition, open ended, *becoming*'. The becoming that emerges from the liminal state affords some kind of resolution in which new status, ways of being come to pass. But, as Björn Thomassen (2009) notes, liminality is not an explanation and there is no certainty of outcome. Uncertainty has always tinged liminality with danger – the initiand may not survive the initiation, the possibility that the purpose of the process will not come to fruition. Since the referendum of 2016, although Article 50 was triggered by Parliament to signal the UK's withdrawal from the EU, the process that followed with the negotiation of the withdrawal agreement was fraught with difficulties which led to an impasse in Parliament. So, while on the one hand, the process of departure was being processed, there was no guarantee that the UK would depart the EU – indeed the departure date was set back three times.

Since December 2019 and the Conservative Party's national election victory, the date for leaving was definitively set for 31 January 2020. However, this by no means is the end of the matter; negotiations with the EU on what post-UK membership relations will look like are on-going and as such Brexit 'has not got done' by 1 February 2020 and still no one knows what exactly 'it' will look like or mean.[8] In addition, the idea that the UK may at some future point rejoin the EU has been floated. For example, Labour leadership contender Jess Phillips suggested that a future Labour government might campaign on such a basis, although she did later modify her remarks.[9] And, it is not just within the physical geographical entity of the UK that individuals or groups may be harbouring hopes that the UK will one day rejoin the EU. What of the expatriates who have made a life elsewhere in Europe? Jeremy MacClancy's (2019) insightful work with expatriates in Alicante, Spain, has shown that members of these communities have organised anti-Brexit groups. For expatriates, their situation is perilous as their position in terms of rights and residency are still to be agreed in negotiations. As Anne Wesemann (2016) pointed out, 'It is entirely up to the national government of each European Union member state to decide how individuals will gain and lose nationality'. Further, as MacClancy (2019: 377) highlights, the plight of the expatriates was not met with 'sympathetic interest from their Spanish neighbours', rather they faced questions such as 'When are you going home?'.[10]

With any expression of sentiments to rejoin the EU coupled with the lingering uncertainty around the shape of a post-Brexit UK, if Brexit were to be likened to the liminality of the ritual process with clearly defined starts, middles and ends, the comparison fails, because currently it seems that there is no end point at which the country will emerge into the promised land or faraway where life will be so much better as a result of withdrawing from the EU. Thomassen (2009: 22) discusses the idea of the 'permanentization of liminality', which he parallels with Turner's (1969) institutionalisation of liminality. He argues that 'without reintegration liminality is pure danger. Hence, relating to crisis periods of larger societies where the social drama has no foregone conclusion, the question becomes: how is the liminal period dealt with, and how (if at all) is it ended?' (Thomassen, 2009: 22).

This book does not seek to answer the question of where all of this will end, or even if it will end, but rather it uses tourism as a lens through which some of the issues that arise as part of Brexit can be examined.

Chapter Outlines

Tom Selwyn's Foreword has already introduced the chapters, exploring the ways in which the book adds to an understanding of the liminality of Brexit as well as identifying some of the key themes to emerge from individual contributions. He also points the way towards what a future research agenda might look like based on the ideas laid down by the contributors to this book.

This introduction compliments the Foreword by exploring the concept of liminality in relation to both tourism and Brexit. It has also suggested how some of the thinking and language of discussion, by both Leavers and Remainers, in the Brexit debate can be couched in terms of magic. The order of the individual chapter contributions proceeds by setting out some contextual perspectives (Chapters 2 and 3) and then moving to specific examples found both in the UK and continental Europe (Chapters 4 to 9). From here, we move to the wider world in terms of the Commonwealth countries (Chapter 10), and then to a focus on a specific example of an overseas territory (Chapter 11). Chapters 12 and 13 direct us more to the future regarding what new travel arrangements between the UK and EU might look like and how a post-Brexit geopolitical world might take shape. The book ends with an Afterword that reflects tourism and Brexit considering the circumstances that have arisen at the time of the book going into production, that of the COVID-19 global pandemic.

In Chapter 2, Alexandra Pimor outlines the mythical origins of Europe. She notes that the EU of today is a political manifestation of the dream of European unity for everlasting peace, noting that the Union was not just about a politico-legal and economic entity but also about fostering a spiritual community. She discusses the spiritual dimension of European integration, fostered through European mobility, travel and tourism, and considers the ramifications and significance of Brexit for the emerging cultural picture of a newer European consciousness.

The focus of Chapter 3 by Catherine Palmer is on questions concerning identity such as 'who am I?' and 'where do I belong?'. She argues that for the UK, Brexit represents the biggest challenge since the Second World War to how the individuals, nations and communities that make up the UK understand and respond to the above questions. The chapter explores what the shifting sands brought about by Brexit might tell us about understandings of identity and of 'Britishness' at this particular moment in time.

Continuing with the theme of identity, Vivian Gornik in Chapter 4 discusses the link between heritage and national identity in a contemporary post-devolution, post-Brexit Britain, where the hegemony of national identities is constantly challenged and up for debate. Using examples from ethnographic fieldwork in Glastonbury, Somerset and Tintagel, Cornwall, she asks if heritage can reflect a contemporary, multicultural British reality. Or is the 'present-centeredness' of heritage too limited by factors like tourism to be a proactive source of diversity and inclusivity discourses?

In Chapter 5, Jennifer Frost and Warwick Frost also pay attention to a form of heritage in the shape of the long history of royal events that function to stimulate national pride and allegiance (Laing & Frost, 2018). The focus of their chapter is to examine how royal events and tourism have been subtly refashioned to accommodate and strengthen new international relationships in the post-Brexit era. They argue that the wedding of Prince Harry to Meghan Markle in 2018 served to shift attention away from Europe to the United States and the Commonwealth. As with previous royal weddings, Harry and Meghan's was a media event that served to convey the importance and enduring qualities of British traditions.

Sabina Owsianowska and Magdalena Banaszkiewcz also make use of media in Chapter 6 in the form of the discourses of tourism promotional materials that inform travels between the UK and Central and Eastern Europe at the time of Brexit. The case studies they use illuminate the concept of 'Britishness' as a key metaphor to understanding the dynamics of identity in a world that is constantly transforming. They note that many of the issues that have informed the Brexit debate go beyond issues of travel, but also note that contemporary tourism and its narratives play a part in people and place promotion that has a role in reinterpreting history and building and strengthening group identity.

Staying with media, but in a completely different setting, in Chapter 7, Mark Casey discusses two highly successful British fictional television programmes which focused on the lives and behaviour of British tourists on Spain's Costa del Sol: *Duty Free* (1984–1986) and *Benidorm* (2007–2018). Drawing on these programmes, Casey theorises the role of nationality and social class in creating a 'them' and 'us' binary between the British tourists and the local population. He argues that the themes present in *Duty Free* and *Benidorm* such as racism, social class, a dislike of Spanish culture and a belief in the superiority of 'the British' echo themes to be found in the campaign to leave the European Union.

In Chapter 8, Hazel Andrews considers the campaign rhetoric used by Vote Leave in the run up to the referendum. As she notes, the campaign used the term 'taking back control' as its main slogan to persuade the electorate to vote against remaining in the EU. Inherent in the taking back of control was the promise of greater freedom for the UK to be in control of its own affairs. The ideas of losing control and freedom infuse many holiday fantasies. Drawing on recent writing in social anthropology about freedom in practice (Lino e Silva & Wardle, 2017), this chapter examines ideas of control and freedom in the Mallorcan resorts of Magaluf and Palmanova.

In Chapter 9, Anne Storch and Nico Nassenstein take us back to Spain with another visit to the island of Mallorca where there are large communities of West African migrants who participate in the mass tourism industry. Many are street vendors selling an array of goods from sunglasses to illegal substances. They are dependent on the annual influx of German and British tourists to earn a living. The chapter, based on ethnographic fieldwork, investigates how these polyglot players perceive and estimate the consequences of Brexit and the linguistic and socioeconomic strategies they use to cope with the potentially decreasing numbers of British tourists.

In Chapter 10, the focus shifts away from Europe to think more about the issues relating to the Commonwealth countries. In this chapter, Marcus Stephenson and Shaun Goldfinch speculate on the UK's future relationship with the Commonwealth nations and how the Commonwealth can be reimagined considering Brexit, especially in terms of international tourism. The chapter also explores two themes found in previous chapters – identity and heritage. Among the questions they ask are 'can heritage representations be constructive rather than a reflection on former core–periphery relations and the hegemonising impact of British colonial governance?' and 'how can the tourism industry, tourism mobility and heritage tourism powerfully reflect a shift towards a revitalised Commonwealth identity based on an Anglophone future – as opposed to a European identity?'.

In Chapter 11, Maria Amoamo considers the reconstitution of the core–periphery in terms of identity by focusing on the implications of Brexit for the UK's 14 overseas territories, most of which are islands. She considers trade, security, financial aid, access, mobility, identity and tourism development. Amoamo argues that Brexit represents a 'core–periphery' shift in the mainly dyadic relationship with the metropolitan or colonial power. The chapter builds on extant literature in tourism studies regarding the concept of 'core–periphery' in general, and more specifically in the study of subnational islands as places of differentiated development processes using a case study of Pitcairn Island.

In Chapter 12, Victoria Hooton takes a socio-legal approach to analysing the concept of 'associate EU citizenship', a proposed legal mechanism which would allow UK citizens, post-Brexit, to opt-in to EU citizenship, and therefore the legal framework that allows for free movement. Her discussion highlights that for many the stripping of EU citizenship creates not only a loss of fundamental rights, but also a loss of European identity that has developed through legal integration, shared history and common values. The potential for associate citizenship to fill that void is assessed and the chapter concludes that the EU citizenship framework caters for specific types of mobility, mainly associated with educated, skilled or wealthy citizens. While associate citizenship will protect fundamental free movement rights for UK citizens fitting this description, the framework could be perceived as divorced from the values attached to EU citizenship's symbolic status.

Chapter 13 by Reza Masoudi discusses issues of globalisation, claiming that Brexit signifies the end of globalisation as it is currently understood and invites the development of a new geography in which the UK itself may well be refigured. He draws parallels with the outcome of the 1979 referendum in Iran that led to the establishment of the country's Islamic Republic in highlighting that the exact implications of the people's votes in both cases was not fully known or understood. The key idea of this chapter is to explain that such referendums are not merely about establishing a new political makeup, but rather the formation of a new geography.

Last word

So much is unknown and cannot be divined about the future of the UK following 31 January 2020. As the country travels in a new direction, the new order of things will materialise in cultural practices of which both travel and tourism are expressions. Border watching has never seemed more important.

Notes

(1) In their reflections on the implications of Brexit 'in the structuration of geographic thought and practice', Boyle *et al.* (2018) cite several academic analyses of the causes of Brexit.
(2) It was/is also connected with internal divisions within the Conservative Party – see Robert Shrimsley's (2018) report in the *Financial Times*. See https://www.ft.com/content/0dee56c0-fdfa-11e8-ac00-57a2a826423e (accessed January 2020).
(3) See also Andrews and Roberts (2012).
(4) By this, I mean that it did not return the majority government she was seeking and meant there was a need to rely on support from the DUP.
(5) See https://www.theguardian.com/uk-news/2014/oct/26/british-towns-swamped-immigrants-michael-fallon-eu (accessed December 2019).
(6) See https://www.newyorker.com/magazine/2019/03/25/the-magical-thinking-around-brexit (accessed December 2019).

(7) See https://www.theguardian.com/commentisfree/2019/apr/12/brexit-halloween-rees-mogg-fear-far-right? CMP.
(8) For an understanding of the possible implications of what Brexit means for the UK, see Brendan Vickers' (2019) thoughts from the perspective of political economy. As the author himself notes, his ideas are based on what was the current thinking at his time of writing in March 2018.
(9) See https://www.independent.co.uk/news/uk/politics/labour-leadership-jess-phillips-eu-brexit-remain-referendum-a9271801.html (accessed January 2020).
(10) Some students and friends of mine from other EU countries reported similar sentiments as those expressed by some of the Spanish towards the British in Alicante when, post-Brexit, they were faced with questions in the UK such as, 'why are you still here?'.

References

Andrews, H. (2000) Consuming hospitality on holiday. In C. Lashley and A. Morrison (eds) *In Search of Hospitality: Theoretical Perspectives and Debates* (pp. 235–254). Oxford: Butterworth Heinemann.
Andrews, H. (2009) Tourism as a 'moment of being'. *Suomen Antropologi* 34 (2), 5–21.
Andrews, H. (2011) *The British on Holiday. Nation, Identity & Consumption*. Bristol: Channel View Publications.
Andrews, H. (2014) The enchantment of violence: Tales from the Balearics. In H. Andrews (ed.) *Tourism and Violence* (pp. 46–68). Farnham: Ashgate.
Andrews, H. (2017) Becoming through tourism. *Suomen Antropologi* 42 (1), 31–44.
Andrews, H. (2019) With Brexit uncertainty heating up, is the tourism industry cooling down? See https://ukandeu.ac.uk/with-brexit-uncertainty-heating-up-is-the-tourism-industry-cooling-down/ (accessed 20 January 2020).
Andrews, H. and Roberts, L. (eds) (2012) *Liminal Landscapes. Travel, Experience and Spaces In-Between*. London: Routledge.
Appiah, K.A. (2016) Lecture 4 'Culture'. BBC Radio 4 Reith Lecture 'Mistaken Identities: Creed, Country, Color, Culture'. Broadcast Tuesday 8 November.
Bialasiewicz, L. (2009) Europe as/at the border: Trieste and the meaning of Europe. *Social & Cultural Geography* 10 (3), 319–336.
Boyle, M., Paddison, R. and Shirlow, P. (2018) Introducing 'Brexit Geographies': Five provocations. *Space and Polity* 22 (2), 97–110.
Brown, G. (2016) We need a Brexit deal that heals the north–south divide. See https://www.theguardian.com/commentisfree/2016/nov/08/gordon-brown-brexit-deal-north-south-divide-power-regions-britain (accessed November 2016).
Bruner, E. (1991) Transformation of self in tourism. *Annals of Tourism Research* 18 (2), 238–250.
Evans, G. (2017) Brexit Britain: Why we are all postindustrial now. *American Ethnologist* 44 (2), 215–219.
Gardner, A. (2017) Brexit, boundaries and imperial identities: A comparative view. *Journal of Social Archaeology* 17 (1), 3–26.
Gardner, A. and Harrison, R. (2017) Brexit, archaeology and heritage: Reflections and agendas. *Papers from the Institute of Archaeology* 27 (1), 1–6.
Gennep, A. van (1960 [1909]) *The Rites of Passage*. Chicago, IL: Chicago University Press.
Graburn, N.H.H. (1976) Tourism: The sacred journey. In V. Smith (ed.) *Hosts and Guests: The Anthropology of Tourism* (2nd edn; pp. 21–36). Oxford: Blackwell.
Gusterson, H. (2017) From Brexit to Trump: Anthropology and rise of nationalist populism. *American Ethnologist* 44 (2), 209–214.
Hobolt, S.B. (2016) The Brexit vote: A divided nation, a divided continent. *Journal of European Public Policy* 23 (9), 1259–1277.

Huang, W.J., Xiao, H. and Wang, S. (2018) Airports as liminal space. *Annals of Tourism Research* 70, 1–13.

Hutt, M. (1996) Looking for Shangri-la: From Hilton to Lāmichhāne. In T. Selwyn (ed.) *The Tourist Image: Myths and Myth Making in Tourism* (pp. 49–60). Chichester: John Wiley and Sons.

Jackson, M. (2013) *Lifeworlds. Essays in Existential Anthropology.* Chicago, IL: Chicago University Press.

Knight, D.M. (2017) Anxiety and cosmopolitan futures: Brexit and Scotland. *American Ethnologist* 44 (2), 237–242.

Kumar, K. (2003) *The Making of English National Identity.* Cambridge: Cambridge University Press.

Laing, J. and Frost, W. (2018) *Royal Events: Rituals, Innovations, Meanings.* London: Routledge.

Lashley, C. and Morrison, A. (eds) (2000) *In Search of Hospitality: Theoretical Perspectives and Debates.* Oxford: Butterworth Heinemann.

Lino e Silva, M. and Wardle, H. (2017) Introduction. Testing freedom. In M. Lino e Silva and H. Wardle (eds) *Freedom in Practice. Governance, Autonomy and Liberty in the Everyday* (pp. 1–33). London: Routledge.

MacClancy, J. (2019) Before and beyond Brexit: Political dimensions of UK lifestyle migration. *Journal of the Royal Anthropological Institute (N.S.)* 25 (2), 368–389.

Malinowski, B. (1948 [1954]) *Magic, Science and Religion and Other Essays.* New York: Anchor Books.

Mauss, M. (1954) *The Gift: Forms and Functions of Exchange in Archaic Societies.* London: Routledge.

Moeran, B. (2016) Performing the EU referendum: A view from afar. *Anthropology Today* 32 (6), 1–2.

Pagden, A. (2002) Europe: Conceptualizing a continent. In A. Pagden (ed.) *The Idea of Europe. From Antiquity to the European Union* (pp. 33–54). Cambridge: Cambridge University Press.

Picard, M. (2010) Tropical island gardens and formations of modernity. In J. Scott and T. Selwyn (eds) *Thinking Through Tourism* (pp. 139–160). Oxford: Berg.

Picard, M. (2011) *Tourism, Magic and Modernity: Cultivating the Human Garden.* Oxford: Berghahn.

Pritchard, A. and Morgan, N. (2006) Hotel Babylon? Exploring hotels as liminal sites of transition and transgression. *Tourism Management* 27 (5), 762–772.

Reeves, M. (2016) Democracy on speed. *Social Anthropology* 24 (4), 479–480.

Searle, J.R. (1998) *Mind, Language and Society: Philosophy in the Real World.* New York: Basic Books.

Selwyn, T. (1996) Introduction. In T. Selwyn (ed.) *The Tourist Image: Myths and Myth Making in Tourism* (pp. 1–32). Chichester: John Wiley and Sons.

Selwyn, T. (2007) The political economy of enchantment: Formations in the anthropology of tourism. *Suomen Antropologi* 32 (2), 48–70.

Selwyn, T. (2010) The tourist as a juggler in a hall of mirrors: Looking through images at the self. In E. Waterton and S. Watson (eds) *Culture, Heritage and Representation: Perspectives on Visuality and the Past* (pp. 195–214). Farnham: Ashgate.

Sheldrick, G. (2020) Britain's £26BN Brexit tourism boom. *Daily Express*, 3 January, p. 1, 4.

Shields, R. (1991) *Places on the Margin: Alternative Geographies of Modernity.* London: Routledge.

Shore, C. (1993) Inventing the 'People's Europe': Critical approaches to European community 'cultural policy'. *Man (N.S.)* 28 (4), 779–800.

St John, G. (2008) Victor Turner and contemporary cultural performance: An introduction. In G. St John (ed.) *Victor Turner and Contemporary Cultural Performance* (pp. 1–37). Oxford: Berghahn.

Stein, F. (2016) Anthropology, Brexit and xenophobia in Europe. See https://politicalandlegalanthro.org/2016/06/28/anthropology-brexit-and-xenophobia-in-europe/ (accessed April 2017).

Thomassen, B. (2009) Uses and meanings of liminality. *International Political Anthropology* 2 (1), 5–28.

Tourism Alliance (2016) EU referendum impacts on the UK tourism industry. See https://www.tourismalliance.com/ (accessed September 2016).

Turner, V. (1969) *The Ritual Process*. Chicago, IL: Aldine.

Vickers, B. (2019) Implications of Brexit. In T.M. Shaw, L.C. Mahrenbach, R. Modi and X. Yi-Chong (eds) *The Palgrave Handbook of Contemporary International Political Economy* (pp. 283–299). London: Palgrave.

Wesemann, A. (2016) Hope for UK nationals living abroad after Brexit. *The Conversation*. See https://theconversation.com/hope-for-uk-nationals-living-abroad-after-brexit-61871 (accessed November 2016).

World Tourism Organisation (1980) Manilla Declaration on World Tourism. See https://www.univeur.org/cuebc/downloads/PDF%20carte/65.%20Manila.PDF (accessed 20 January 2020).

2 'Travel broadens the mind and enriches the soul': Exploring the Significance of Tourism and Brexit on the Spiritual Dimension of European Integration

Alexandra Pimor

Introduction

This chapter, premised on the conception of European integration, Brexit and tourism as phenomena, argues that tourism is an integral vehicle of the spiritual dimension of European integration, and posits that Brexit reveals a lacuna in the current debates on the future of Europe which persist in being led by economic objectives to the detriment of the cultural and spiritual unity of the Union.

European integration has predominantly been framed as an economic endeavour, a narrative served to the British nation since it voted to join the European Union (EU) in 1972. Yet, while the Single Market was the first concrete step towards *de facto* solidarity, the vision of the founding fathers was more ambitiously holistic and encompassed a united Europe for peace, prosperity, defence and spiritual community (de Rougemont, 1965). While the pragmatist ideological path to the European construction favouring peace through prosperity prevailed, spiritual community-building received less attention and is thus little understood.

Tourism, defined by the United Nations World Tourism Organisation (WTO) as 'a social, cultural and economic phenomenon', is deemed 'a major unifying force in Europe and contribution to integration' (Steppova, 2002: para. 15). In supporting the peoples of Europe to expand their experiences and understanding of each other through diverse exchanges (economic, political, cultural, etc.), mobility, travel and tourism are essential factors in the development of a European consciousness shaped by common values and shared principles. The adage that 'travel broadens the mind' is therefore at the heart of the multifaceted European integration process. Nevertheless, the political and economic slant remains strongly prevalent in the debates surrounding the future of Europe, Brexit and its consequences. A hard Brexit suggests a complete exit from any type of customs union with a direct impact on UK–EU migration patterns, which will be affected by the elevation of barriers to the free movement of persons, whether for work, residence or tourism purposes. All the protections and advantages afforded by EU citizenship status (contained in Articles 20–24 of the Treaty on the Functioning of the European Union) will be lost to the British people, including voting and consular rights. However, Brexit also has significance for European spiritual unity.

This chapter is divided into two parts: the first contextualises the European venture as a spiritual journey that is supported by structural elements of *de facto* solidarity such as the Single Market; while the second part explores the implications and challenges of the Brexit phenomenon both for the British people and for the future of the EU and integration process.

The European Spiritual Journey and Dimension of Integration

The idea of Europe

A creation of the mind

Before becoming a continent, Europe was first and foremost a construct of the mind, a convoluted idea that has become an *ideal* inspiring ambitions of conquest seemingly worth fighting for. Unifying the diverse territories, kingdoms and nations of Europe in one form or another has been a long-winded process spanning centuries of empire-building, characterised by an oft violent and pugnacious tenacity to unite through subjugation and domination. Since the 1300s, over 180 plans to unite Europe have been concocted (Lascaris, 2008), some having left an indelible mark on shaping the continent, most notably the Roman Empire, Christendom and the Third Reich Nazi regime. This journey reflects the mythological story of Europa, the Asian princess of yore after whom the continent was named, and whose tale of abduction, rape and metamorphosis epitomises the essence of European history (Pagden, 2002). Europe's mythological origins account for her metamorphosing ability, her protean nature (forever transforming herself and shape-shifting in accordance with the flow of history, or *her* story) leading to a new imagined reality that is the EU – a construct founded on a myriad of debated myths (Hansen & Williams, 1999; Kølvraa, 2016; Probst, 2003).

Still, in renouncing the belligerent psychology of national dominance, today's unification of European member states is the only (relatively) successful instance of nation-states that have come together voluntarily and peacefully in a spirit of conciliation and cooperation, to create something greater than the sum of their national parts. This required vision, fortitude and a compassionate psychology that Europe can heal and transcend the conflicts of her past, to realise a future filled with hope for the peaceful unification and good life of her peoples. Founding fathers Schuman, Adenauer and de Gasperi, in particular, drew inspiration from their Christian-Democrat ideology, fuelled by their cosmopolitan upbringing, experiences of the Second World War and tenets of their faith, to define the lines of engagement for the future of a united Europe. They envisioned the unification in a spirit of forgiveness, solidarity and equality (Pimor, 2015). The spirit of reconciliation is at the heart of European integration (Guisan, 2013: 15); and from this historical perspective of personal and national transcendence, it can be said that the 'European experience transmits ... a vision of optimism' (Guisan, 2013: 59).

A spiritual adventure

For all the above reasons, Europe truly is a *spiritual* adventure (Delors, 2011) which, given the French origin of the modern European unification project (i.e. the Schuman plan), refers to the French meaning of the term, that is *esprit*. *L'esprit* is the human intellect or the mind, understood as the seat of human thoughts and ideas. While *l'esprit*, and so the spirit, may relate to the incorporeal part of being human, a distinction is to be made between the religious meaning and the wider, fluid and secular understanding of spirit/ual/ity. Although Christianity is an integral part of European history, and Christian values of forgiveness and reconciliation inspired alternative international politics to combat the *état d'esprit contraire* or the nationalism that resulted in the enduring fragmentation of Europe (Schuman, 1953), the founding fathers observed and upheld the principle of *laïcité*, that is the separation between state and church, politics and religion. Consequently, the focus of this chapter is on the spiritual dimension, i.e. relating to the sociocultural, psychological, philosophical and even religious influences that have informed the architects, builders and gatekeepers of European unity, and not on the confessional, denominational, dogmatic and institutional religious aspects of European integration.

The idea of Europa thus began in the minds of her peoples – her philosophers, scholars, artists and religious and political leaders, as did her wars, a seemingly inevitable consequence of the ideal to unite the countries of her continent (de Rougemont, 1961). To paraphrase the preamble to the UNESCO Constitution (1945), since wars began in the minds of men, it was in their minds that the defences of peace had to be constructed. Peace is the aim of a united Europe (Article 3 TEU): peace from war, peace for and among the peoples of Europe, and the peaceful enjoyment of a good, healthy and prosperous life for their well-being

(Pimor, 2017). European integration is thus a historically rooted enterprise that fulfils a moral necessity that is both essential to the common good of humanity and to the revitalisation of Europe, for which spiritual rebirth is imperative (McCauliff, [1950] 2012: 449).

It is in the minds of its founding fathers that the dream of a united Europe became a possibility, a vision that galvanised governments into peaceful cooperative action. The founding fathers consciously harnessed that vision of European unity under an internationally agreed plan for collective and supranational association, for the common good and the pursuit of common objectives. They succeeded in manifesting this vision into reality, a feat of peaceful engineering that was only ever possible as they adopted a new frame of mind that underlined the intentions and behaviours of European political actors and national governments.

Incremental integration process

Multifaceted European unity

The European unification venture is based on and reflects four key historical themes of unity: peace, defence, prosperity and spiritual community, whereby peace is facilitated through the higher authority of supranational institutions in order to counter any claim to hegemony; a common defence is coordinated to guarantee Europe's security (particularly against outside threats); prosperity for all is organised on a continental scale; and the spiritual community reflects the European unity of traditional, creative and diverse cultures, and is supported by a common legal framework (de Rougemont, 1965).

In the years preceding the creation of the first European communities, the founding fathers emphasised the importance of spiritual unity in the making of Europe. Europe was ready to seek unity through cooperation between willing partners to ensure 'her material and moral recovery' and establish peace and security among her states and peoples (Schuman, 1949). The first step was the establishment of the Council of Europe for the defence, protection and promotion of human rights. In a speech given in London in 1949, Robert Schuman declared that the council's statute was laying down 'the foundations of *a spiritual* and political cooperation from which there will arise the *European spirit*, the promise of a broad and lasting supranational nation'. The unity of Europe would thus be achievable through the development of the European spirit, which in practice would both support and be supported by supranational institutions and economic integration. Inspired by a 'consciousness of European unity, common destiny, obstacles and tasks to be fulfilled', the creation of supranational institutions for a European union was deemed the means by which this spiritual unity, this spirit of service to Europe and the world would be facilitated and reinforced through the permanent cooperation of its peoples (Schuman, 1953).

Therefore, the European unification process would not be linear or uniform, it would be built on concrete steps, 'on reality' and 'by the placing in common of her resources' (Schuman, 1950). It would be realised through its various communities and start with *de facto* (concrete) solidarity since, before Europe's spiritual unity could be tended, she had to regain her material independence post-1945 (Schuman, 1953).

Proposals for European political and defence communities were presented and vetoed (by the French government) due to concerns about safeguarding national sovereignties. Therefore, the less threatening and more conceivable economic route was selected as the first concrete step towards European integration and *de facto* solidarity. Although all sorts of intermediate steps could have been taken with great benefit, including a thinking revolution (e.g. the abandonment of absolute national sovereignty to a federal system for the members of a collaborative international organisation such as the Community), it was conceded that an economic union was an extreme though ideal solution (Spaak, 1950: 94, 97–98). Schuman argued that the economic 'functional' approach had been chosen for practical reasons, which were three-fold. Firstly, 'it seemed wiser to begin with integration in a restricted, technical sector of national life'. Secondly, this option allowed for the unification project to move rapidly while 'catching the public imagination and winning over the doubters and reluctant'. Thirdly, economic integration was less susceptible to political controversy. As such, the spirit and will of union would have to be expressed externally in a community of action and institutions that would give them real effectiveness (Schuman, 1953).

De facto solidarity: The economic route underpinning efforts for spiritual unity

Given the destruction and devastation yielded by the war that ravaged Europe and changed the world, the European Coal and Steel Community (ECSC) created in 1951 was first and foremost a means to prevent such reoccurrence. With a 'never again' mentality underlying its conception, the aim of the ECSC was to promote peace and cooperation between France and Germany by pooling their coal and steel resources (means of production for the machinery of war) together under a supranational regulatory framework. In so doing, the ECSC fostered a common-interest approach which neutralised the national self-interested tendencies that could lead to armed conflict.

The Treaty of Rome in 1957 gave rise to the European Economic Community (EEC), which was initially a regeneration project created to rebuild a decimated European economy in order to sustain its peoples and to protect Western Europe against the advance of totalitarian communism. The EEC marked the beginning of the Customs Union and its four fundamental freedoms of movement, including goods, capital, services and workers/persons. By 1992, the Single Market was established creating one European economic territory.

The Common Market was created as the economic means to rebuild a devastated Europe, progressively ensuring unhindered mobility of workers/people, goods, capital and services within its internal borders, since the evolution of Europe has historically been manifested through the movements and migration of its peoples (Favell, 2014). Mobility and travel are inherent freedoms, rights and occurrences within the construct of the EU, which fulfil both economic and cultural functions in its integration process. Indeed, 'economic integration is not an objective in itself, its rationale is to serve higher objectives, both of an economic and a political nature', including economic welfare, peace and security, democracy and the protection of human rights (Molle, 2006: 4); and of course, spiritual and cultural unity.

The evolution from the free movement of workers (under the EEC) to the free movement of persons in 1992, with the establishment of European citizenship under the Treaty of Maastricht, reinforces the observation that economic integration is but an aspect of the broader concept of European integration; a means to an end that provides the physical underpinning of the Union's *raison d'être* to unite not only the states but also the peoples of Europe in a spirit of solidarity and within a supranational setting.

Therefore, free movement of individuals also means the free movement of *ideas* between European countries for economic, political, scientific and cultural purposes, which is a prerequisite condition to unity (Schuman, 1953); for the spiritual journey of unification means freedom of thought and action in overcoming the obstacles to the union 'which are first and foremost obstacles of the mind, and not of fact' (de Rougemont, 1948). The four fundamental mobility freedoms are the cornerstone of the European integration process through the Single Market structure, which is 'now the place where Europeans can enjoy a unique diversity of culture, ideas and traditions in a Union covering four million square kilometres. It is where they have forged lifelong bonds with other Europeans and can travel, study and work across national borders without changing currency' (European Commission, 2017).

The Impact of Tourism and Brexit on European Integration

Peace, integration and tourism

The powerful myth of peace through tourism

To reiterate, the aim of European unity is peace based on reconciliation and cooperation, facilitated by the Single Market (no internal borders) for the free movement of European citizens – whether to pursue work or education, or simply to travel and tour the territories of the EU's member states. As Mark Twain (1869) remarked, 'travel is fatal to prejudice, bigotry, and narrow-mindedness' and putatively offers a means for European peoples to form a mutual understanding of each other and bridge their differences, and thus practice peace through tourism (Bechmann Pedersen, 2017). As such, tourism constitutes a recognised element of European integration – both economically and culturally (Diaconou, 2015).

The concept of peace through tourism is highly debatable yet still worth considering in the context of European narratives of integration. Peace through tourism essentially relies on contact theory, whereby in fostering contact between peoples, tourism creates an environment for mutual understanding, sympathy and friendship, which in turn leads to a reduction in conflicts (Bechmann Pedersen, 2017: 30). Tourism is seen as a vital force for peace and social change (Goh, 1988), and deemed a vehicle for bridging distances among peoples of different races, ethnicities, religious denominations and socioeconomic backgrounds, as it promotes mutual trust and respect through the rich and diverse tapestry of human cultures and environments (D'Amore, 1988: 270). However, although it is agreed that tourism may, through cross-cultural interaction, contribute to opening minds and widening people's views by changing negative stereotypes and promoting good will between peoples (Pratt & Liu, 2015), it appears to be more of a political sentiment than an objectively observable and measurable phenomenon (Bechmann Pedersen, 2017).

Therefore, while the idea that tourism promotes peace has been researched, there is no clear empirical evidence that it does promote peace. Recent literature presents mixed views on the relationship between peace and tourism, essentially showing the idea of peace through tourism as a theoretical debate, often based on an unfounded hypothesis as the product of political rhetoric and the travel industry proselytising (Bechmann Pedersen, 2017; Farmaki, 2017; Pratt & Liu, 2015).

Still, the concept is an influential myth that has historically been perpetuated through international and European tourism policy framing over decades. Throughout the 1940s and 1950s, peace was claimed to be served by cross-national travel and friendships; a post-Stalinist Soviet Union became Western-tourists friendly; diplomats and travel industry agents attending an international conference on tourism in Prague confidently declared this would 'contribute to the expansion of international tourism… and to the strengthening… of the ideal of peace'. In 1964, the United Nations issued its Recommendations on International Travel and Tourism, stating that tourism promotes international goodwill, understanding and the preservation of peace between peoples. In 1967, the UN General Assembly launched the International Tourist Year with the motto 'Tourism, passport to peace' (Bechmann Pedersen, 2017: 33–34). In its Manila Declaration in 1980, the WTO declared that it was 'convinced that world tourism can be a vital force for world peace and can provide the moral and intellectual basis for international understanding and interdependence' – an axiom echoed by a Tourism Industry Association of Canada (TIAC) resolution adopted in support of the WTO 1986 International Year of Peace. Peace through tourism remains an enduring myth, as evidenced by the *International Handbook on Tourism and Peace* (Wohlmuther & Wintersteiner, 2014) published in partnership with the UNWTO.

Overall, while it may be difficult to convincingly conclude whether tourism leads to peace, it is more widely recognised that peace is an element that promotes tourism rather than the other way around, and that tourism not only benefits from but depends on peace more than the opposite (Pratt & Liu, 2015). It is also recognised that tourism can lead to political, social and economic transformation in peoples' lives by providing a space and forum for narratives of reconciliation and peace to emerge (Farmaki, 2017). Nevertheless, and more cynically perhaps, the questionable assumption that tourism is a vehicle of peace can be rebutted by the observation that this interwar internationalist discourse was adopted by the tourist industry to legitimatise the growth of mass tourism (Bechmann Pedersen, 2017: 35). Either way, whether tourism contributes to peace or whether peace leads to tourism seems to be a 'chicken or egg' dilemma that, in the context of European integration may be moot, for tourism (i.e. travel and mobility) is both an integral element and inherent product of European integration.

Tourism policy in the EU: A convoluted narrative

The founding fathers, and Schuman in particular, argued that spiritual unity would be achieved through different means, i.e. cultural exchange among the peoples of Europe. Therefore, the broad concept of spiritual unity is to be understood as including the cultural, philosophical and knowledge dimension of European consciousness. This exchange was advocated through promoting the free movement of persons, ideas, arts, etc. Travel (unhindered and even supported) and therefore free mobility arguably constitute a key element of the spiritual integration process of the Union.

However, Schuman also believed that the spiritual unification of Europe could only be achieved through free accords and not through a supranational will. The approach for unifying Europe would thus require variation and flexibility as 'the supranational principle should not apply to the cultural as there should not be any standardisation of the spirit' (Schuman, 1953). Consequently, tourism (understood as a cultural phenomenon) initially fell outside the competences of the Union, although since 2009 the economic significance of the travel and tourism industry has been guaranteed a legal basis under Article 6(d) and Article 195 of the Treaty on the Functioning of the European Union (TFEU).

Tourism comes under the Union's *supporting* competences, in that the EU can only act in the field of tourism policy to support, coordinate or complement the action of EU countries (Art. 6 TFEU). In other words, while tourism remains within the remit of national governments' strategies and policy-making, the Union can adopt certain measures for the development of a European framework on tourism. Under Article 195 TFEU, the Union has competence to promote competitiveness in the tourist sector, encourage cooperation between member states and develop an integrated and thus coherent approach to include tourism within the European body of policies in other areas. However, Article 6 TFEU not only provides a legal framework of collective action on tourism at the European level, but it also ensures that there are limits to the EU's competence to intervene. Indeed, the Union cannot adopt harmonising legislation in that field and must respect the principles of proportionality and subsidiarity (Art. 5 TFEU) when intervening to fulfil treaty objectives – the former demands that the Union's action must be proportionate to the aims pursued to achieve the objectives; while the latter allows the EU to intervene only if a member state's actions are insufficient and treaty objectives can be better achieved through Union action.

The inclusion of tourism in a constitutional founding treaty of the EU denotes a concern by member states and Union institutions to adopt a more concerted collective approach in fostering the development of tourism in Europe and tackling the challenges presented by the emerging economic, financial and environmental crises since 2008. In a communication entitled 'Europe, the world's No 1 tourist destination – a new political framework for tourism in Europe', the European Commission (2010) qualifies tourism as a major economic activity and industry that must also be addressed in light of sustainability and ethical principles, given its impact on European cultural and natural heritage. In 2015, the European Commission branded tourism as the third largest socioeconomic activity in the EU (after the trade and distribution, and construction sectors) with a positive impact on economic growth and employment (Juul, 2015: 5).

EU tourism policy is thus framed in an economic narrative and the term 'tourism industry', though debatable, is widely used and refers to tourism in relation to financial services, gross domestic product (GDP) growth, employment and labour force, the consumption of goods and services and wider economic sectors' contributions (European Parliament, 2019). This development in the European tourism policy narrative has arguably been reinforced by the economic account of European integration that became dominant with the objective to pursue unification through the completion of the Single Market.

In this framing, tourism ceases to be a cultural phenomenon and culture instead becomes an incidental consequence or appendage of the tourism industry (e.g. cultural tourism, see Mousavi *et al.* [2016] for a definition). The European Commission (2010: 6) notes that tourism only *indirectly* 'helps to strengthen the feeling of European citizenship by encouraging contacts and exchanges between citizens, regardless of differences in language, culture or traditions', a far cry from the peace through tourism mantra of the early European reconciliation project, or the founding fathers' vision of European cultural and spiritual unity through free movement of people and ideas. This is problematic when terms such as *benefit tourism*, *health* or *education tourism* emerge in the European integration discourse and are addressed from the cold eye of economic objectives and immigration policy analysis, rather than the compassionate and *solidaire* sentiments and expressions of spiritual unity.

It must be said that, while economic integration essentially defined the Union as a problem-solving, instrumental, functional and thus limited organisation, the economic narrative traditionally left citizens aloof, as long as the Union was working and providing adequate solutions to market-based issues. However, the economic, refugee and asylum crises have engendered more existential questioning about the EU and its relevance. De Gapseri had warned at the onset of European unification that an economic technocratic Union 'without a higher political will invigorated by a central body in which the wills of nations come together, are fully expressed and come alive in a higher framework... may seem cold and lifeless – it could even at times appear a superfluous and even oppressive extravagance' (Paparella, 2012: 126–127).

The rise in nationalist, far-right and populist politics across the Union, the very sentiments the founding fathers of European unity laboured and designed the communities against, seems to bring the Union full circle. It can be said that life is not a linear process, but a circular experience, and that we revisit certain situations either because we have not learnt from the event or because we are ready to discover another perspective and gain deeper wisdom from that new occurrence. So, what does Brexit tell us about the European unification and spiritual adventure?

Brexit: A spiritual crisis?

The economic and material impact of Brexit

In this chapter, Brexit is perceived as a phenomenon, i.e. a situation that is observed to exist and may be apprehended from various perspectives. Still, it is impossible to confidently explain or even define the meaning, nature and full range of the implications of Brexit in the long term – both for the UK and the EU. It is, and will continue to be, analysed from diverse disciplinary prisms (political, legal, sociological, economic and so on) and is likely to remain elusive to any attempt at providing a unified analysis – much like Puchala's (1971) pachydermesque allegory of the field of European integration theories.

From an economic perspective, since membership of the EU was sold to the British people as associating with member states in a mere common market, it is also on economic arguments that the case for Brexit was promoted with, as commented in the media, the (arguably misguided) assumption that Britain retains the (former) glory and significance of the British Empire (Younge, 2018). While there are plenty of news reports on the British prime minister repeatedly claiming that the UK will fare better economically outside the EU (under WTO rules even) and will conclude many trade agreements with various countries (e.g. the United States, Canada, Australia and other Commonwealth members), the reality is that those claims are often rebuked by business and trade organisations and experts.

The consequences of Brexit on the economic life of the UK and its people have been explored in a briefing paper to the House of Commons 2019. The impact of Brexit on UK tourism is far-ranging. It is noted that, since the positive results to the 2016 referendum on Brexit, there has been a fall in the number of European visitors to the UK, from 72% in 2016 to 64% in March 2019 (Foley & Rhodes, 2019: 5). Still, EU tourists accounted for the majority of visitors to the UK in 2018 (71%) and 49% of visitors' spending (Foley & Rhodes, 2019: 8).

The challenges identified by the briefing paper include the potential loss of EU funding to tourist areas and destinations; loss of consumer protection rights (e.g. roaming fees, package holiday insolvency remedies, European health insurance services); and restrictions and higher fees on air travel services. In addition, 'the tourism industries are considerably more reliant on the employment of people from other EU countries than the rest of the economy. So, if Brexit resulted in lower levels of EU migration to the UK and fewer EU workers, this would have a disproportionate effect on the UK tourism sector' (Foley & Rhodes, 2019: 22).

The wider significance of Brexit

However, in the economic narrative of European integration and now that of Brexit, a crucial point is missing: how Brexit affects the cultural and spiritual unity of the European integration venture; and what it means for the British people and for EU citizens, respectively. Indeed, while the UK may leave the European construct of the Union, it is still part of the European continent and thus still *European*. Although, given the historical UK–EU lukewarm relations (Schweiger, 2007), it is valid to wonder whether Brexit was an inevitability that has come to pass and 'the UK was destined to leave, regardless of which side won the referendum' (Smith [2017] as reported in Haagedoorn's blog).

Despite advocating for a kind of united states of Europe, Churchill (1946) declined to enjoin the UK to the European unification project, and in so doing created an ambiguous relationship expressed by the preposition 'with' but not 'of', whereby Britain is *with* and *linked* to Europe, but not *of* and *comprised* in it. Notwithstanding the special bond between the UK and Europe, there was initially no desire to lock Britain in what was expected to be a federal union, since Britain had its own Commonwealth and Empire (Risse, 2015: 82). When the UK eventually joined the EU, it was confronted with the realisation that it had lost an opportunity to influence the moulding of European institutions, and faced both an established institutional setting and key constitutional principles underpinning the Union legal order, i.e. the doctrines of direct effect and primacy of EU law over conflicting domestic law developed by the European Court of Justice in case 26/62 *Van Gen den Loos* and case 6/64 *Costa v ENEL*. From that late arrival and the perception that these EU doctrines led to the curtailment of parliamentary sovereignty following the House of Lords judgement in *Factortame I* [1990] 3 CMLR 1, Euroscepticism became embedded in British politics, in both left and right political discourse. Although that did not stop the UK from subsequently becoming one of the key three players in shaping European policy (along with France and Germany).

Hence, Brexit appears to be a tale of British equivocation about its Union membership and was thus also campaigned on political and cultural grounds. It was won on claims of European democratic deficit; Union institutions' technocratic distance from its citizens; British parliamentary sovereignty claw-back; national political and immigration policy independence; and financial and economic benefits redirected back to UK public services. The referendum was as much about the EU as it was about British national and party politics, the result of a failed attempt to bridge divisions within the Tory Party over the European question. Brexit was sold as an exit from the Union construct, though without details, clear explanations of what that would entail or how the future shape of UK–EU relations would be (Smith, 2017).

Yet, Brexit is not solely about socioeconomic and political ramifications. It is, on the one hand, the moral revocation of the reciprocal pledge between the UK and its sister nations to unite in a spirit of solidarity and collective supranational action in the practice of peace. On the other hand, Brexit is an indictment of the gatekeepers of the united European endeavour locked into a political, financial and economic state of mind that has forgone the vision of spiritual and cultural unity dreamt by Europe's founding fathers. In elevating physical and material barriers between the UK and the EU, the Brexit phenomenon is erecting borders in the minds of European peoples who forged a spiritual bond over the past six decades of weaving a mosaic European family united in diversity. The Brexit phenomenon is evidence that the dream of a united Europe as a spiritual endeavour must be reclaimed.

Conclusion

While for many years it was argued that Europeans were indifferent about the Union, media frenzy and European citizens' responses over Brexit have revealed that belonging (or not) to the EU can generate the highest of emotional responses. Certainly, as a nation, 54% of the British hold a favourable view of the EU (Wike *et al.*, 2019). The Eurobarometer (2019 – Spring) findings further show that over 50% of British citizens view themselves as EU citizens and enjoy the benefits of free movement. Therefore, Brexit may paradoxically prove to be a hindrance and an unexpected opportunity for the furtherance of European solidarity and spiritual unity. Just as it may be generating a political–cultural schism, it may also contribute to a spiritual ascension for British/EU citizens in their lived and perceived experience of the EU.

The EU has often – and somewhat imprecisely – been penned a 'peace project'. However, a project suggests an end once the goal is achieved, yet peace is not a static state of mind or a fixed state of being, it is a practice contingent on the evolutive meaning of 'peace'. The EU might have been a reconciliation project for its initial period of construction, but today has become a way of life (as suggested by the newly formed 2019 European Commission portfolio on the European Way of Life), anchored in a practice of peace through cooperation, consensus-building and collective decision-making.

Brexit raises questions for the future of European integration beyond its economic image, yet the European Commission (2017) offers more economic scenarios on the future of Europe that fail to address the spiritual *raison d'être* of continued European unification. In addition, traditional European integration theories are not equipped to provide a convincing analysis of the Brexit phenomenon and other rumoured exits (Kostakopoulou, 2017); and do not offer a framework for observation and discussion on the significance of European spiritual and cultural unity, which is admittedly not easily observable. Given that the EU is a spiritual adventure, it is time to direct our attention towards the *spiritual and moral heritage* of the Union as enshrined in the preamble to the European Union Charter of Fundamental Rights. Brexit is therefore not the end, but the beginning of a new quest for the Union's soul, and marks another step on the transformative spiritual journey of European unity.

References

Bechmann Pedersen, S. (2017) Peace through tourism: A brief history of a popular catchphrase. In I.M. Andrén (ed.) *Cultural Borders and European Integration* (pp. 29–38). Göteborg: Centrum for Europeaforskning.

Churchill, W. (1946) Speech delivered at the University of Zurich, 19 September 1946. See https://rm.coe.int/16806981f3 (accessed 1 July 2020).

D'Amore, L.J. (1988) Tourism: A vital force for peace. Research notes and reports. *Annals of Tourism Research* 15, 269–283.

Delors, J. (2011) 'L'Europe, une aventure spirituelle'. Discours dans l'Institut Catholique de Paris, 24 November 2011. *Transversalités* 2012/3 (23), 119–132.

de Rougemont, D. (1948) *L'Europe en jeu*. Neuchâtel: Éditions de la Baconnière. See https://www.cvce.eu/content/publication/2008/1/18/ff3d3e3a-0f5b-41bf-961a-0678 22bb65ee/publishable_en.pdf

de Rougemont, D. (1961) *Vingt-huit siècles d'Europe*. Poitiers: Payot.

de Rougemont, D. (1965) Europe unites. A lecture on 'The History of the Ideal for a United Europe'. *The Meaning of Europe* (trans. A. Braley). London: Sidgwick & Jackson.

Diaconou, M. (2015) Towards an integrated approach to cultural heritage for Europe. European Parliament Committee on Culture and Education. Report 2014/2149(INI).

Eurobarometer (2019) Standard Eurobarometer 91 – the key indicators – United Kingdom. Spring 2019. See file:///C:/Users/alexp/Downloads/eb_91_fact_uk_en.pdf (accessed 1 July 2020).

European Commission (2010) Europe, the world's No 1 tourist destination – a new political framework for tourism in Europe. Communication to the European Parliament, the Council, the European Economic and Social Committee and the Committee of the Regions. COM/2010/0352 final. See https://eur-lex.europa.eu/legal-content/EN/TXT/?uri=celex:52010DC0352 (accessed 1 July 2020).

European Commission (2017) White Paper on the Future of Europe. Reflections and scenarios for the EU27 by 2025. COM (2017) 2025. See https://ec.europa.eu/commission/sites/beta-olitical/files/white_paper_on_the_future_of_europe_en.pdf (accessed 1 July 2020).

European Parliament (2019) Tourism. See https://www.europarl.europa.eu/ftu/pdf/en/FTU_3.4.12.pdf (accessed 1 July 2020).

Farmaki, A. (2017) The tourism and peace nexus. *Tourism Management* 59, 528–540.

Favell, A. (2014) The fourth freedom: Theories of migration and mobilities in 'neo-liberal' Europe. *European Journal of Social Theory* 17 (3), 275–289. doi: 10.1177/1368431014530926.

Foley, N. and Rhodes, C. (2019) Tourism: Statistics and policy. *House of Commons Library Briefing Paper*. Number 06022, 24 September 2019.

Goh, P. (1988) Tourism research: Expanding boundaries. *Travel and Tourism Research Association 19th Annual Conference*, Montreal, Quebec, Canada, 19–23 June (pp. 199–208). Salt Lake City, UT: University of Utah.

Guisan, C. (2013) *A Political Theory of Identity in European Integration – Memory and Policies*. Abingdon: Routledge.

Haagedoorn, S. (2017) Blog post: Summary of Dr Julie Smith's lecture the historical evolution of UK–EU relations. European Council Library Blog. 2 June. See https://www.consilium.europa.eu/fr/documents-publications/library/library-blog/posts/the-historical-evolution-of-eu-uk-relations/ (accessed 6 December 2019).

Hansen, L. and Williams, M.C. (1999) The myths of Europe: Legitimacy, community and the 'crisis' of the EU. *Journal of Common Market Studies* 37 (2), 233–249.

Juul, M. (2015) Tourism and the European Union. Recent trends and policy developments. European Parliamentary Research Service (EPRS). See https://www.europarl.europa.eu/RegData/etudes/IDAN/2015/568343/EPRS_IDA(2015)568343_EN.pdf (accessed 1 July 2020).

Kølvraa, C. (2016) European fantasies: On the EU's political myths and the affective potential of utopian imaginaries for European identity. *JCMS Special Issue: Another Theory of Possible: Dissident Voices in Theorising Europe* 54 (1), 169–184.

Kostakopoulou, D. (2017) What fractures political unions? Failed federations, Brexit and the importance of political commitment. *European Law Review* 42 (3), 339–352.

Lascaris, A. (2008) Healing Europe. In J. Juhant and B. Zalec (eds) *Surviving Globalization: The Uneasy Gift of Interdependence. Theological Perspectives on Religion in Europe Bulletin ET* – 19 (2008/1) 56–61

McCauliff, C. (1950) Union in Europe: Constitutional philosophy and the Schuman declaration, May 9, 1950. *Columbia Journal of European Law* 18, 441–472.

Molle, W. (2006) *The Economic of European Integration: Theory, Practice, Policy.* Aldershot: Ashgate.

Mousavi, S.S., Dorati, N. and Moradiahari, F. (2016) Defining cultural tourism. *International Conference on Civil, Architecture and Sustainable Development (CASD-2016),* 1–2 December, London.

Pagden, A. (2002) *Europe: Conceptualising a Continent. The Idea of Europe.* Cambridge: Cambridge University Press.

Paparella, E. (2012) *Europa: An Idea and a Journey – Essays on the Origins of the EU's Cultural Identity and Its Economic-Political Crisis.* Bloomington, IN: Xlibris.

Pimor, A. (2015) The unbearable elusiveness of the European Union's spiritual heritage. *Journal for the Study of Spirituality* 5 (1), 33–46.

Pimor, A. (2017) Solidarity was a founding principle of European unity – it must remain so. *The Conversation* UK, 24 March. See https://theconversation.com/solidarity-was-a-founding-principle-of-european-unity-it-must-remain-so-74580 (accessed 6 December 2019).

Pratt, S. and Liu, A. (2015) Does tourism really lead to peace? A global view. *International Journal of Tourism Research* 18 (1), 82–90.

Probst, L. (2003) Founding myths in Europe and the role of the Holocaust. *New German Critique* 90, 45–58.

Puchala, D.J. (1971) Of blind men, elephants and international integration. *Journal of Common Market Studies (JCMS)* 10 (3), 267–284.

Risse, T. (2015) *A Community of Europeans?: Transnational Identities and Public Spheres.* Ithaca, NY: Cornell University Press.

Schuman, R. (1949) Statement on 5 May 1949, at the signing of the Statute of the Council of Europe in London, the French Foreign Minister Robert Schuman calls for a revival of the European spirit and hails the dawn of new political cooperation in Europe. See https://www.cvce.eu/obj/statement_by_robert_schuman_london_5_may_1949-en-97217713-8cb6-4679-a6ce-8bd77660bd17.html (accessed 6 December 2019).

Schuman, R. (1950) Statement proposing the creation of the ECSC, declared on 9 Math 1950, as then French Foreign Minister in the Salon de l'Horloge at the Quai d'Orsay in Paris. See https://europa.eu/european-union/about-eu/symbols/europe-day/schuman-declaration_en (accessed 6 December 2019).

Schuman, R. (1953) Comment traduire publiquement l'unité historique et culturelle des européens? Exposé introductif présenté à la cinquième séance de la Table Ronde de l'Europe, organisé par le Conseil de l'Europe à Rome 13–16 octobre 1953 sur le thème de *L'Unité spirituelle et culturelle de l'Europe et la mission des européens dans le monde contemporains.* Conseil de l'Europe. TR (53) 3 Strasbourg 20 Septembre 1953. See http://www.coe.int/t/dgal/dit/ilcd/Archives/selection/Schuman/TableRonde53_fr.pdf (accessed 1 July 2019).

Schweiger, C. (2007) The reluctant European: Britain and European integration since 1945. In C. Schweiger (ed.) Britain, Germany and the Future of the European Union. *New Perspectives in German Studies* (pp. 14–42). London: Palgrave Macmillan.

Smith, J. (2017) *The UK's Journeys Into and Out of the EU. Destinations Unknown.* London: Routledge.

Spaak, P.H. (1950) The integration of Europe: Dreams and realities. *Foreign Affairs* 29 (1), 94–100.

Steppova, V. (2002) Tapping Europe's tourism potential. Committee on Economic Affairs and Development. Report. Doc. 9461. Council of Europe (9 May 2002).

Twain, M. (1869) *The Innocents Abroad.* Ebook. See https://www.gutenberg.org/files/3176/3176-h/3176-h.htm (accessed 6 December 2019).

UNESCO Constitution (1945) Preamble, signed 16 November 1945 at the ECO/CONF/29. See http://www.unesco.org/education/pdf/UNESCO_E.pdf (accessed 6 December 2019).

Wike, R., Poushter, J., Silver, L. and Cornibert, S. (2019) European Public Opinion Three Decades After the Fall of Communism: Most embrace democracy and the EU, but many worry about the political and economic future. Pew Research Centre. See https://www.pewresearch.org/global/wp-content/uploads/sites/2/2019/10/Pew-Research-Center-Value-of-Europe-report-FINAL-UPDATED.pdf (accessed 6 December 2019).

Wohlmuther, C. and Wintersteiner, W. (eds) (2014) *International Handbook on Tourism and Peace*. Klagenfurt: Centre for Peace Research and Peace Education of the Klagenfurt University in cooperation with World Tourism Organization (UNWTO). See https://www.e-unwto.org/doi/pdf/10.18111/9783854357131 (accessed 1 July 2020).

World Tourism Organisation (WTO) (1980) Manilla Declaration on World Tourism. See https://www.univeur.org/cuebc/downloads/PDF%20carte/65.%20Manila.PDF (accessed 6 December 2019).

Younge, G. (2018) Britain's imperial fantasies have given us Brexit. *The Guardian*. 3 September. See https://www.theguardian.com/commentisfree/2018/feb/03/imperial-fantasies-brexit-theresa-may (accessed 6 December 2019).

3 The Order of Things: Brexit and the Challenge of Identity

Catherine Palmer

Introduction

Questions concerning identity, such as 'who am I' and 'where do I belong', continue to frame and influence the geo-political landscape, a landscape constantly straining and shifting in response to local, regional and world events. For the United Kingdom of Great Britain and Northern Ireland (UK), Brexit represents the biggest peacetime challenge since the Second World War to how the individuals, nations and communities that make up the UK understand and respond to the above questions. This chapter explores what the shifting sands brought about by Brexit might reveal about understandings of identity and of 'Britishness' at this particular moment in time.

In terms of the UK, identity(ies) and 'Britishness' are negotiated in relation to borders and boundaries – actual, symbolic and imaginary – and defined as much by the self as they are by others. How the 'we' are seen by others has been brought to greater prominence through the Brexit process; however, so far, one significant other, that of tourism (and tourists), has not been centre stage when the potential implications of Brexit are discussed. While this may be understandable, a nation that is redefining its economic and political relationship with its closest neighbours and the wider world is also redefining its identity, its internal and external 'face'.

National identity is, of course, a complex and contradictory concept, particularly so for the UK where Great Britain is sold as a destination in its own right and collective references to British identity are not always welcome as they obscure the strongly held affiliations to the countries and regions that comprise the wider UK. Despite the challenges of identity, British identity and Britishness do have popular currency through usage by other nations, different forms of media, by marketing agencies and, of course, tourism and tourists. Hence, my use of British and Britishness is not meant to ignore the complexities inherent in questions concerning identity, it is rather an acknowledgement that this is how the identity of the UK is largely referred to for tourism purposes.

How might an evolving internal 'face' influence the external 'face' presented to the 'you' of tourism? What might this evolving 'face' reveal about the role of identity in shaping the relationship between self, other and the wider world? There are no definitive answers to these questions, they are more thoughts out loud than solvable problems, but they are nonetheless important. Firstly, because they highlight what for me is the message of Brexit, that it is first and foremost about identity and its interlocutors' belonging and self-determination and secondly, because they are good to think with.

Thinking with and through the issues raised by the above questions can shine a light on the relationships and attitudes that shape how individuals in the UK come to understand how the nation is or should be 'ordered'. In saying this, I am making a link between Brexit and a classic concern of social anthropology, which is uncovering how the whole of a society is held together (or torn apart) by beliefs and attitudes about the particular; beliefs associated with particulars such as how houses should be built, about attitudes towards pollution and risk, about the role of food in maintaining the social order and about the significance of a cut finger (Bloch, 1995; Douglas, 2002 [1996]; Gell, 1996; Selwyn, 1980). Brexit has disrupted the taken-for-granted ordering of life from which individuals and groups construct a sense of Britishness. It has destabilised the established order of things on many different levels; however, perhaps the most significant disruption is to long-held understandings of sovereignty and hence power. Does the power to make decisions lie with the people through their involvement in electing politicians to represent them? Or, does it lie with democratically elected politicians who then decide what laws are needed to enable them to represent the people? The referendum has undermined the taken-for-granted order of things based upon understandings of where the power to decide lies, people or Parliament, and this is why the nation has been so divided. Although these questions are not the only explanations for the Brexit vote, they have produced a range of responses from the polemic to the considered (Bogdanor, 2019; Liddle, 2019; O'Toole, 2018). Ultimately, through the referendum it is 'the people' not the political elite deciding the nation's future direction and, depending upon the way someone voted, this future is either feared or viewed as an exciting opportunity. As one journalist commented:

> In the entire Brexit saga, however, the usual order has been reversed. The people 'out there' have understood more quickly than the people in the thick of it what is going on… The message from the sticks to the centre is that everything is about to change. (Moore, 2019: 20)

As the nation has struggled to respond to Brexit, tourism's role is to promote a recognisably British 'face' internally as well as externally. This chapter will reflect upon and respond to the thoughts set out above.

The Challenge of Identity

Any discussion of identity and belonging needs to acknowledge that they are complex conjoined twins frequently contested and debated (Frosh & Baraitser, 2009; Hall, 1996; Hobsbawm, 1992). Identity is also a complex lived experience that can be destabilised by events such as Brexit (Browning, 2018). Complexity is nowhere more evident than when focusing on identity in relation to nations constructed from other highly visible national, regional and ethnic identities such as the United Kingdom of Great Britain and Northern Ireland.

According to the historian, Linda Colley, British national identity was forged in 1707 when the Act of Union, which brought Scotland into a union with England and Wales, established the legal entity that is the UK. This date marks the point at which 'the invention of Britishness' began (Colley, 1994: 1). Since 1707, national cohesion and identity have been threatened by wars and political and constitutional crises, most notably the Napoleonic war with France spanning the 18th and 19th centuries. Although the need to defend the nation and its sense of self from the threat of a French invasion was very real (Colley, 1994), it was not until the Second World War that Britishness as a manifestation of 'what we stand for' was physically attacked at home by the bombing campaigns of Nazi Germany. Brexit resembles an attack on identity from within, a self-inflicted wound in the foundations of Britishness and, as such, it marks a particular moment in the evolution of both an individual and a collective sense of self. Brexit has pulled apart the 'order of things' upon and against which this sense of self is constructed. Brexit is British identity in a moment of crisis (Brown, 2019; Dettmer, 2019; Oliver, 2017; O'Toole, 2019; Storry & Childs, 2017a).

Within the UK, the three countries collectively referred to as Great Britain are England, Scotland and Wales so in one respect there is no such thing as a homogeneous sense of Britishness or British identity for those individuals who primarily identify as English, Scottish or Welsh. Although, of course, many individuals do define their identity as being solely British or a combination of British and English, Scottish or Welsh. In terms of Northern Ireland, identity is complicated by religion, which in turn is deeply embedded in the geography and politics of place. As a result identity (in particular national identity) is contested, at times violently so with, in simplistic terms, unionists defining their identity as British with strong allegiance to the British state and the Crown, while republicans identify with and seek to be part of the wider independent nation of Ireland. For a more detailed exploration of the issues outlined above see Kumar (2010), McCrone and Bechhofer (2015), Nic Craith (2003), Todd (2015) and Storry and Childs (2017b).

The political complexities of the UK are beyond the scope of this chapter but it is important to mention them here because they are indicative of the literal and symbolic borders and boundaries that cut across discussions of identity in relation to the UK generally and Brexit in particular (see Gardiner, 2017; Werbner, 2017; Willett et al., 2019). They are also illustrative of existing complexities and contradictions even before Brexit came out to play. Furthermore, to speak of national identity is to speak of only one type of identity, one expression of who I am because identity can be understood in terms of the self, rather than or as well as the whole; a nation, community or tribe. As such identity can mean different things to different people. It can be expressed in relation to context and situation meaning that self-identity (personhood, sense of self) is invariably a kaleidoscope of other identities that are constantly reshaping themselves in response to any given situation (Hitchcock, 1999; Jenkins, 2014).

Hence, an ethnic or national identity can sit alongside an identity linked to a set of beliefs, a faith or religion; identity can be constructed in terms of political affiliation, in relation to an occupation, race, caste or social class. Gender and sexuality are significant identity(ies); however, these categories are fluid rather than biologically stable since gender and sexuality are socially constructed non-binary namings open to individual interpretation (Butler, 2004; Caplan, 1989; Hall & Bucholtz, 1995; Shrage, 2009). In addition to or instead of these examples, self-identity can be based around membership of social groupings or engagement with leisure activities and has become increasingly experienced digitally via the construction of a digital self (Buckingham, 2008; Murray, 2015; Thumim, 2012). It is no wonder then that McCrone and Bechhofer (2015: 17) define (national) identity as an active doing rather than a static naming, as a verb 'to identify with' rather than a noun, 'In this way, by treating people as agents, we are better able to get at how and why people mobilise national identifications, and for what purpose'.

Although McCrone and Bechhofer were writing before Brexit, their point is prescient given the outpouring of collective and individual angst since 2016. Brexit has influenced how an individual understands his/her sense of self generally and within the context of affiliation to a collective sense of nationness (Balthazar, 2017; Browning, 2018; Johnston, 2018; Wincott et al., 2019). Various commentators and academics have pointed to the emergence of a new identity based on voting preferences as being either a remain or a leave supporter (Clarke et al., 2019; Curtice, 2018). However, such affiliations are unlikely to stand the test of time; they are convenient and temporary labels rather than new identities capable of sustaining a collective sense of nationness for the long term.

Brexit also divided the nation by laying bare how existing fault lines relating to region, class, education and ethnicity, and race influenced the outcome of the referendum (Ford & Goodwin, 2017; Henderson et al., 2016; Namusoke, 2016; Willett et al., 2019). As Henderson et al.'s (2017: 631) analysis of survey data demonstrated 'Brexit was not just made in England... Englishness was also a significant driver of the choice to leave'. However, it is important to remember that such a statement does not fully explain the complexities inherent in the reasons why individuals voted leave (see Clarke *et al.*, 2017).

Interestingly, in reflecting back on her ethnography of British charter tourists to Spain, Andrews (2017) argues that the regional identities and divisions displayed at the time of the research illustrate why the regional voting patterns that brought about Brexit might not be that surprising. Caution is needed, however, in drawing conclusions from opinions expressed and surveys undertaken in the eye of the storm. Time for empirically derived data and reflection derived from distance is needed before any sense of what the whole Brexit moment might reveal about how societies and individuals change, adapt and move forward generally and specifically in relation to identity. Anthropologists have a unique opportunity here as connections and relationships established through ethnography can enable Brexit to be incorporated into existing fieldwork. For example, MacClancy's (2019) work on citizenship and UK lifestyle migration before and beyond Brexit is directly linked to his long-term ethnographic work in the Alicante province of Spain. Likewise, Manley's PhD research on Scottish nationalism is well placed to locate the Brexit moment within a focus on activism in relation to independence (Manley, 2019).

The fact that identity is challenging, contradictory and contested despite Brexit is what makes it such a fascinating aspect of Brexit. Brexit has unsettled seemingly taken-for-granted understandings of 'who I am' and 'where I belong' not just for the individual, but also within family and friendship groups, work and faith communities, and for political and cultural elites. Externally, the relational ties between Great Britain and the Commonwealth, the political association of nations largely drawn from what was once the British Empire caused one commentator to state that Brexit merely highlights how divided the Commonwealth 'family' really is (Namusoke, 2016). Brexit has affected how the nation is seen by others with many news and media outlets expressing consternation at what has happened to a country whose centre of government, frequently referred to as the cradle of democracy has been in such chaos and whose reputation for pragmatism, diplomacy and self-restraint seemingly vanished (Henley, 2019; Howell, 2018).

Although Brexit was largely 'made in England', leading Calhoun (2017: 57) to argue that 'Brexit is an expression of English (more than British) nationalism', it may turn out to be a highly significant moment in the history of identity formation. Clearly, identity is constantly being made and remade in response to events, but Brexit is a very specific event in the timeline of identity formation stretching back to 1707 because, as noted earlier, it is a self-inflicted wound in the foundations of Britishness. Despite the fact that foundations crack and move over time, what matters is that enough people agree to repair, reshape and maintain them because what they reinforce, a collective sense of cohesion or nationness, is deemed worth supporting. With Brexit, a battle is taking place over different versions of what the nation stands for and how it should represent itself internally and externally. It is a battle over what it feels like to belong, what it feels like to be British, as a headline in *The Irish Times* newspaper put it 'It is not just the economy, stupid – Brexit is about belonging' (O'Toole, 2019). The battle for identity and belonging is out in the open, in the home, at work and at play, in the tangible and the virtual communities to which people belong. It is a spectacle for others to observe, a very public breakdown of the order of things understood as Britishness. Where does tourism sit within this reordering of identity? What will a post-Brexit Britishness look like through the eyes of tourism?

Brexit, Identity and Tourism

Clearly, Brexit needs to have happened followed by enough time for a post-Brexit world to be encountered and experienced before a more informed response to these questions can be attempted. Although the UK formally left the European Union (EU) in January 2020, the transition year for agreeing the future relationship has been severely disrupted by the social and economic upheavals caused by the COVID-19 pandemic. Despite these upheavals, in June 2020 the British government ruled out an extension to the transition period meaning that the look and feel of a post-Brexit world remains uncertain. Not least because the consequences of a possible 'no deal' outcome are difficult to comprehend and, as yet, impossible to assess. However, it is possible to think through the existing relationship between tourism and identity generally and as it relates to countries who are or have gone through profound change to reconfigure their identity for tourism purposes.

Generally, tourism as the internal and external face of a nation gathers together the different strands of a nation's identity drawn from such elements as history, myth, systems of governance, music, literature, food, landscape, language and so on and weaves them into a coherent story of nationness capable of representing the nation's identity in ways that differentiate it from other nations. This differentiation is marketed as a product and a particular type of experience, where identity becomes a commercialised brand. There is a wealth of literature on the ways in which tourism works to create a particular type of identity including the consequences for places and for people – tourists and locals (for example de Jong, 2020; Light, 2000; Palmer, 2005). Much has also been written about tourism as a tool for nation-building after conflict, so the significance of tourism for reconfiguring people and place is already well recognised (for example Amujo & Otubanjo, 2012; Causevic & Lynch, 2011; Vitic & Ringer, 2008). However, the UK is not emerging from commonly understood versions of conflict such as wars, power struggles over territory, cultural legitimacy or resources.

The fabric of the nation has not been attacked, damaged or destroyed; armies have not been mobilised and citizens have not died fighting to defend the nation; territory has not been ceded or borders redrawn. The nation's cartography remains intact. Yes, people marched in protest for and against Brexit; heated arguments occurred online; Brexit was discussed at work, at home and in the pub; and there was constant media reporting *ad nauseam*. The positive and negative implications of Brexit for the economy generally, for trade and for employment and welfare remain ever present. The media are never short of individuals and lobby groups willing to talk about the issues affecting specific sectors such as business, finance, transport, the National Health Service, manufacturing and so on.

Voices from within the UK and from within this country's international visitor economies questioning the implications for tourism are being heard, but understandably the topic of conversation is dominated by economics, employment, transport and practicalities such as travel insurance, the loss of reciprocal arrangements for medical care, visa free travel and mobile phone charges (ABTA, 2019; Mason, 2019; Mills, 2018; Minihane, 2019; Sanders, 2016). As one travel website succinctly put it: 'How will Brexit affect your holiday?' (holiday extras, 2019). The likely impact on currency exchange rates and the practical and legal implications for travelling to the UK generally and in terms of the potential consequences of the UK leaving without an agreement, the 'no-deal' scenario, are major concerns (Dobruszkes, 2019; Papí Ferrando *et al.*, 2018; Trend, 2017; VisitBritain, 2019). Even though the government's concerted push to prepare for a no-deal Brexit might represent a kind of certainty for leave supporters, the unknown practical and legal consequences of such an outcome for overseas travellers to and from the UK are profoundly unsettling. Will planes still fly? Will cars be stuck at border crossings? How will ships and cruise liners be processed in Southampton, Dover or Glasgow? Will my passport still 'work'?

Such questions and concerns are understandable given the debilitating effect of uncertainty. Discussions of how Brexit might influence the collective face the nation presents to other countries through tourism are largely absent from the public discourse. Yet, nationness profoundly influences individual and collective understandings of belonging which in turn maintain the collective 'we' that enables a nation to exist; as Miller (2000: 32) argues, one function of national identity is that it serves to create a sense of 'solidarity, such that people feel themselves to be members of an overarching community.... Nationality is *de facto* the main source of such solidarity'. 'Solidarity' is rarely, if ever, as uniformly solid as it might appear on the surface, aptly demonstrated by the Brexit crisis, described as 'at best a group of nations under a single sovereign, now being pulled apart by Europe's gravitational field' (Scruton, 2016). The possible implications and consequences of Brexit's gravitational pull for the nation's well-being in relation to its sense of self is of equal significance as those relating to economics, finance and transport. In fact, and I would argue, it is of greater significance because economics, finance, systems of government and so on are underpinned by the values and beliefs that support a nation's sense of self. However, the citizens and seat of government have been divided over and actively questioned these values and beliefs; questioned the 'face' presented internally and externally. Meaning that potential visitors to the UK might question their taken-for-granted assumptions and understandings of who and what the country *is*.

In terms of the visitor experience, then, in one sense the well-established, highly visible attractions, landmarks and events employed to construct and promote the nation's identity for tourism purposes are not going to change. Great Britain's 'brand' identity as promoted by the main regional tourism agencies, *VisitBritain*, *VisitScotland* and *Visit Wales* will continue to refer to and publicise the regions, cities and destinations in much the same way as they currently do. Post-Brexit, the landscape and major attractions will be as they are now; Great Britain will still have a monarchy and a Parliament, the key sites associated with cities such as London, Cardiff and Edinburgh will continue to be recognisable as Great Britain. The country's tourism product based largely upon a cultural narrative drawn from history, heritage, ancient monuments, a constitutional monarchy – in effect what the country is known for will not be replaced by anything different post-Brexit. Although, at some point, the Brexit moment will become part of the history that informs the construction and understanding of identity.

For all intents and purposes, the London Eye will continue to revolve, Shakespeare's Stratford will not move and the William Wallace Monument in Stirling will still stand as a physical and symbolic reminder of Scotland's fight for independence from English rule. The geographic borders and boundaries will not change, these will continue to identify the landmass that is the UK-Great Britain – notwithstanding decisions that need to be made about the border in Northern Ireland – Hadrian's Wall will not fall down, the Scottish Highlands will not disappear nor the Giant's Causeway in Northern Ireland be swallowed up by the Atlantic Ocean. Sterling will remain the currency, and all the pomp, pageantry and national events associated with 'being British' such as the national anthem, Trooping the Colour, sporting events and so on will continue. For all intents and purposes the 'things' of tourism will be ordered as they are now. What Brexit has influenced is how the nation sees itself and how it is seen by current and potential tourists. The tourist version of Great Britain may not be changing but taken-for-granted assumptions of what Britishness looks like internally and externally are affected by Brexit and, given the turmoil the country is in, how the national 'we' is viewed is unlikely to settle down until well into a post-Brexit phase.

Flatman (2017) does speculate about whether Brexit will lead to changes in the legislation that protects the heritage, particularly in terms of *Historic England*, the governmental organisation charged with protecting England's historic environment. He argues that funding, conservation priorities and strategies may change enabling less orthodox sites such as those associated with the nation's colonial past, with slavery and acts of repression, to play a more substantial part in constructing the national self-narrative. However, why any legislative changes cannot be made irrespective of Brexit is unclear.

Likewise, the need to provide a more diverse and inclusive picture of the past as heritage has been obvious for years. The EU has not prevented those governmental and non-governmental agencies responsible for the historic environment, and from which a tourism product is constructed, from acknowledging and including alternative stories of the past. Although there are exceptions and things are changing, it is not an issue of EU or UK legislation but one of representation in terms of who decides what counts as heritage.

So, if the fabric from which tourism builds a nation's identity does not change, then what will tourists, domestic, regional and international encounter? What, if anything will be different as a result of the Brexit challenge to the 'order of things'? Clearly, the legal and political entity that is the UK will be different because the country will no longer be part of or bound to the wider legal and political entity that is Europe. The ordering that defines the nation as independent within Europe will have profoundly shifted to one where the nation is independent in so far as a sovereign Parliament outside Europe allows (Bogdanor, 2019). Whether or not such a change occurs, the meaning and experience of being who 'we' are will have changed, affecting what it feels like to belong; what it feels like to be me, us, we; and what it feels like to encounter you, they, them. The 'feel' of a place for tourists can be subtle, contradictory and ephemeral, but this does not mean that it is not a significant part of a tourist's experience of place. Uncertainty brings with it the possibility of the new, the different, the transformative and for the individual this can be positive or negative depending upon how he/she voted in 2016. Tourists will wonder what has changed, what is new. Who are these people if they are not part of Europe?

One response to this is to be found in *VisitBritain*'s 2018 global marketing campaign to attract overseas visitors called 'I Travel for', comprising a series of eight short films and story-telling drawing on the history and contemporary culture of Great Britain (VisitBritain, 2018a). The films focus on categories such as stories, local flavour, culture, discoveries, relaxation, adventure, the unexpected and fun. The launch film ended with the words 'so, whatever you travel for. Find Your GREAT

Britain' (VisitBritain, 2018b). The GREAT Britain to be found, the face that greets the tourist and which represents the nation's identity, might look the same on the surface, but underneath it will be held together by a different experience of the order of things.

Conclusion

Brexit is 'an illustration of identity dynamics in the long term' (Gardiner, 2017: 3). Only time will tell if it is seen as a pivotal moment in the transformation of national identity, but it certainly feels like it at the time of writing. As Gamble (2018: 1215) asks in relation to the long-term political implications of Brexit, 'Will it come to be seen as a watershed in the political development of the UK, or only an episode in a continuous process of evolution?'. To my mind, a watershed is part of a continuous process of evolution, a process that propels individuals and collective groupings such as nations to adapt to a changing environment. Whether this change is forced and feared or actively sought and welcomed, it is a necessary part of the evolution of being who I am, who we are. Tourism will respond and adapt to change in whatever form as the consequences of Brexit for understandings of identity and belonging become clearer.

The message of Brexit is one thing, but the significance of Brexit is that it represents a challenge to established understandings of how the nation is or should be 'ordered'. This challenge manifests as a struggle between a largely remain-orientated Parliament and a democratically derived vote to leave, 'Brexit is now coming about, not because government or Parliament want it, but because the people want it' (Bogdanor, 2019: 110). This struggle reveals different understandings of the foundational order underpinning the nation's identity, an order that relates to power, to who is actually in charge of the nation's internal and external face. Who decides, people or Parliament? As a result of Brexit, a new principle has entered the language of Britishness, 'the principle of the sovereignty of the people' (Bogdanor, 2019: 110). Although in terms of authority this is not the only principle highlighted by Brexit, since Boris Johnson's attempt to suspend Parliament in 2019 reveals the complex layering of authority at work within the British constitution. This attempt was about the government claiming authority over Parliament. A government led by a prime minister who at the time had been elected by his party rather than by 'the people' via a general election. Although prime ministers are not directly elected by UK citizens and other prime ministers have come to power in the same way, arguments about legitimacy are always arguments about where the power lies to disrupt, change or reconfirm the 'order of things'. As Balandier (1972: 35) states in the classic text *Political Anthropology*, 'The function of power, then, is to defend society against its own weaknesses, to keep it in good "order", one might say'. In terms of Brexit, who decides what counts as being 'in good order' lies at the heart of the UK's political storm because changing the order of things is changing the order that defines who we are.

In this sense, Brexit is a challenge to the prevailing political and social mechanisms of authority that enable a version of the nation to take centre stage. A version controlled by those in power at any given time and capable of glossing over or drowning out alternative versions. The nation that emerges post-Brexit will have been transformed. How this nation will be viewed by tourists in the long term is as yet unknown, but who the nation is, who the British people are will be seen through the prism of Brexit for some time to come. Boris Johnson, as one of the leaders of Vote Leave, stated at the end of a televised debate during the referendum campaign that if people voted to leave then 23 June 2016, the date of the referendum, 'could be our country's Independence Day'. Johnson was clearly meaning independence from Europe, but one of the unintended consequences of Brexit might be independence in terms of 'the people's' ability to exercise control over their elected representatives, control over Parliament. However, getting what you wish for is not always a good thing and in terms of tourism, independent travel can be risky. As one travel website states 'Solo travel can be the ultimate in self-indulgence' (Smarter Travel, 2019). And finally, just in case you are wondering, I voted remain.

References

ABTA (2019) ABTA says it is critical Government delivers the right skills and labour policies for a post-Brexit Britain. See https://www.abta.com/news/abta-says-it-critical-government-delivers-right-skills-and-labour-policies-post-brexit-britain (accessed 10 July 2019).

Amujo, O.C. and Otubanjo, O. (2012) Leveraging rebranding of 'unattractive' nation brands to stimulate post-disaster tourism. *Tourist Studies* 12 (1), 87–105.

Andrews, H. (2017) Touring the regions: (Dis)uniting the Kingdom on holiday. *Journeys* 18 (10), 79–106.

Balandier, G. (1972) *Political Anthropology*. London: Penguin.

Balthazar, A.C. (2017) Made in Britain. Brexit, teacups, and the materiality of the nation. *American Ethnologist* 44 (2), 220–224.

Bloch, M. (1995) The resurrection of the house amongst the Zafimaniry. In J. Carsten and S. Hugh-Jones (eds) *About the House: Levi-Strauss and Beyond* (pp. 69–83). Cambridge: Cambridge University Press.

Bogdanor, V. (2019) *Beyond Brexit. Towards a British Constitution*. London: I.B. Tauris.

Brown, G. (2019) Brexit has sparked a crisis of identity and what it means to be British. *Channel News Asia*, 7 June. See https://www.channelnewsasia.com/news/commentary/brexit-british-values-crisis-of-identity-what-it-means-to-be-eu-11602104 (accessed 27 June 2019).

Browning, C. (2018) Brexit, existential anxiety and ontological (in)security). *European Security* 27 (3), 336–355.

Buckingham, D. (ed.) (2008) *Youth, Identity and Digital Media*. Cambridge, MA: MIT.

Butler, J. (2004) *Undoing Gender*. London: Routledge.

Calhoun, C. (2017) Populism, nationalism and Brexit. In W. Outhwaite (ed.) *Brexit: Sociological Responses* (pp. 57–76). London: Anthem Press.

Caplan, P. (ed.) (1989) *The Cultural Construction of Sexuality*. London: Routledge.
Causevic, S. and Lynch, P. (2011) Phoenix tourism: Post-conflict tourism role. *Annals of Tourism Research* 38 (3), 780–800.
Clarke, S., Kommenda, N. and Lewis, P. (2019) Leavers v remainers: How Britain's tribes compare. Survey reveals surprising consensus – but big divisions exist over identity, culture and gender. *The Guardian*, 2 May. See https://www.theguardian.com/world/ng-interactive/2019/may/02/leavers-v-remainers-how-britains-tribes-compare (accessed 9 May 2019).
Clarke, H.D., Goodwin, M. and Whiteley, P. (2017) *Why Britain Voted to Leave the European Union*. Cambridge: Cambridge University Press.
Colley, L. (1994) *Britons: Forging the Nation 1707–1837*. London: Pimlico.
Curtice, J. (2018) The emotional legacy of Brexit: How Britain has become a country of 'remainers and leavers'. WUKT-EU Briefing Paper. See https://ukandeu.ac.uk/research-papers/the-emotional-legacy-of-brexit-how-britain-has-become-a-country-of-remainers-and-leavers/ (accessed 9 May 2019).
de Jong, A. (2020) Embodying dyke on bike: Motorcycling, travel and the politics of belonging on-the-move. In C. Palmer and H. Andrews (eds) *Tourism and Embodiment* (pp. 107–120). Abingdon: Routledge.
Dettmer, J. (2019) UK struggling with post-Brexit identity crisis. *Voice of America*, 2 January. See https://www.voanews.com/europe/uk-struggling-post-brexit-identity-crisis (accessed 10 July 2019).
Dobruszkes, F. (2019) Air services at risk: The threat of a hard Brexit at the airport level. *Economy and Space A* 51 (1), 3–7.
Douglas, M. (2002 [1996]) *Natural Symbols* (2nd edn). London: Routledge.
Flatman, J. (2017) Identity, value and protection: The role of statutory heritage. *The Historic Environment: Policy and Practice* 8 (3), 181–187.
Ford, R. and Goodwin, M. (2017) Britain after Brexit. A nation divided. *Journal of Democracy* 28 (1), 17–30.
Frosh, S. and Baraitser, L. (2009) Goodbye to identity? In A. Elliot and P. Du Guy (eds) *Identity in Question* (pp. 158–169). London: Sage.
Gamble, A. (2018) Taking back control: The political implications of Brexit. *Journal of European Public Policy* 25 (8), 1215–1232.
Gardiner, A. (2017) Brexit, boundaries and imperial identities: A comparative view. *Journal of Social Archaeology* 17 (1), 3–26.
Gell, A. (1996) Reflections on a cut finger: Taboo in the Umeda conception of the self. In M. Jackson (ed.) *Things as They are: New Directions in Phenomenological Anthropology* (pp. 115–127). Bloomington and Indianapolis, IN: Indiana University Press.
Hall, K. and Bucholtz, M. (1995) *Gender Articulated. Language and the Socially Constructed Self*. New York: Routledge.
Hall, S. (1996) Introduction. Who needs 'identity'? In S. Hall and P. du Guy (eds) *Questions of Cultural Identity* (pp. 1–17). London: Sage.
Henderson, A., Jeffery, C., Lineira, R., Scully, R., Wincott, D. and Wyn Jones, R. (2016) England, Englishness and Brexit. *The Political Quarterly* 87 (2), 187–199.
Henderson, A., Jeffery, C., Wincott, D. and Wyn Jones, R. (2017) How Brexit was made in England. *The British Journal of Politics and International Relations* 19 (4), 631–646.
Henley, J. (2019) UK's reputation for diplomacy, pragmatism and self-restraint has all but gone, say politicians and officials across EU. *The Guardian*, 20 March. See https://www.theguardian.com/politics/2019/mar/20/pathetic-incoherent-chaotic-europes-verdict-on-brexit-shambles (accessed 10 July 2019).
Hitchcock, M. (1999) Tourism and ethnicity: Situational perspectives. *International Journal of Tourism Research* 1 (1), 17–32.
Hobsbawm, E. (1992) *Nations and Nationalism Since 1780* (2nd edn). Cambridge: Cambridge University Press.

holiday extras (2019) How will Brexit affect your holiday? See https://www.holidayextras.co.uk/travel-blog/news/how-will-brexit-affect-travel.html (accessed 27 June 2019).

Howell, D. (2018) The children take over at Westminster. *The Japan Times* See https://www.japantimes.co.jp/opinion/2018/12/17/commentary/world-commentary/children-take-westminster/#.XQy15YUin4c (accessed 21 June 2019).

Jenkins, R. (2014) *Social Identity* (4th edn). Abingdon: Routledge.

Johnston, P. (2018) Brexit is the start of a new argument over national identity. *Public Finance*. See https://www.publicfinance.co.uk/opinion/2018/11/brexit-start-new-argument-over-national-identity (accessed 27 June 2019).

Kumar, K. (2010) Negotiating English identity: Englishness, Britishness and the future of the United Kingdom. *Nations and Nationalism* 16 (3), 469–487.

Liddle, R. (2019) *The Great Betrayal. The True Story of Brexit*. London: Constable.

Light, D. (2000) Gazing on communism: Heritage tourism and post-communist identities in Germany, Hungary and Romania. *Tourism Geographies* 2 (2), 157–176.

MacClancy, J. (2019) Before and beyond Brexit: Political dimensions of UK lifestyle migration. *Journal of the Royal Anthropological Institute (N.S.)* 25, 368–389.

Manley, G. (2019) Scotland's post-referenda futures. *Anthropology Today* 35 (4), 13–17.

Mason, C. (2019) Mobile firms refuse to rule out return of roaming charges after Brexit. *Money Saving Expert*. See https://www.moneysavingexpert.com/news/2019/02/mobile-firms-not-ruling-out-roaming-charges-after-brexit/ (accessed 10 July 2019).

McCrone, D. and Bechhofer, F. (2015) *Understanding National Identity*. Cambridge: Cambridge University Press.

Miller, D. (2000) *Citizenship and National Identity*. Cambridge: Polity Press.

Mills, J. (2018) We could lose the right to urgent healthcare in the EU after Brexit. *METRO*, 28 March. See https://metro.co.uk/2018/03/28/lose-right-urgent-healthcare-eu-brexit-7422986/ (accessed 10 July 2019).

Minihane, J. (2019) How will a no-deal Brexit hit travel in and out of the UK and Europe? *CNN travel*, 16 January. See https://edition.cnn.com/travel/article/how-brexit-will-affect-travel/index.html (accessed 10 July 2019).

Moore, C. (2019) Boris the PM must act decisively in the first 100 days – or it will all end in tears. *The Telegraph* 6 July, 20.

Murray, D. (2015) Notes to Self: The visual culture of selfies in the age of social media. *Consumption Markets & Culture* 18 (6), 490–516.

Namusoke, E. (2016) A divided family: Race, the commonwealth and Brexit. *The Round Table* 105 (5), 463–476.

Nic Craith, M. (2003) *Culture and Identity Politics in Northern Ireland*. Basingstoke: Palgrave MacMillan.

Oliver, T. (2017) A new British identity is key to Brexit's success. So who do we want to be? *The Guardian*, International edition, 16 August. See https://www.theguardian.com/commentisfree/2017/aug/16/british-identity-key-brexit-crisis-negotiations (accessed 27 June 2019).

O'Toole, F. (2018) *Heroic Failure: Brexit and the Politics of Pain*. London: Head of Zeus.

O'Toole, F. (2019) It is not just the economy, stupid – Brexit is about belonging. *The Irish Times*, 22 January See https://www.irishtimes.com/opinion/fintan-o-toole-it-is-not-just-the-economy-stupid-brexit-is-about-belonging-1.3765447 (accessed 10 July 2019).

Palmer, C. (2005) An ethnography of Englishness: Tourism and national identity. *Annals of Tourism Research* 32 (1), 7–27.

Papí Ferrando, J.F., Alfonsi, R., Langer, S. and Troncoso, M. (2018) Research for TRAN Committee – BREXIT: Transport and tourism – the consequences of a no-deal scenario. European Parliament, Policy Department for Structural and Cohesion Policies, Brussels. See http://www.europarl.europa.eu/thinktank/en/document.html?reference=IPOL_STU(2018)617499 (accessed 27 June 2019).

Sanders, R. (2016) The implications of Brexit for the Caribbean's future relationship with Britain and the EU. *The Round Table* 105 (5), 519–529.

Scruton, R. (2016) Who are we? *Prospect*, 14 July. See http://www.prospectmagazine.co.uk/features/who-are-we (accessed 10 July 2019).

Selwyn, T. (1980) The order of men and the order of things: An examination of food transactions in an Indian village. *International Journal of the Sociology of Law* 8, 297–317.

Shrage, L.J. (2009) *'You've Changed': Sex Reassignment and Personal Identity*. Oxford: Oxford University Press.

Smarter Travel (2019) Single travel: Tips for going solo. See https://www.smartertravel.com/single-travel-tips-going-solo/ (accessed 10 July 2019).

Storry, M. and Childs, P. (2017a) Foreword. In M. Storry and P. Childs (eds) *British Cultural Identities* (5th edn; pp. xi–xii). Abingdon: Routledge.

Storry, M. and Childs, P. (eds) (2017b) *British Cultural Identities* (5th edn). Abingdon: Routledge.

Thumim, N. (2012) *Self-Representation and Digital Culture*. Basingstoke: Palgrave-MacMillan.

Todd, J. (2015) Partitioned identities? Everyday national distinctions in Northern Ireland and the Irish state. *Nations and Nationalism* 21 (1), 21–42.

Trend, N. (2017) 10 ways Brexit could hit UK travellers. *The Telegraph* (online). See https://www.telegraph.co.uk/travel/comment/what-would-brexit-mean-for-travellers/ (accessed 21 June 2019).

VisitBritain (2018a) VisitBritain launches new global campaign to boost inbound tourism. See https://www.visitbritain.org/visitbritain-launches-new-global-campaign-boost-inbound-tourism (accessed 21 June 2019).

VisitBritain (2018b) Find Your GREAT Britain. See https://www.visitbritain.com/gb/en/i-travel/stories (accessed 21 June 2019).

VisitBritain (2019) Advice for visitors on travel to Britain in 2019. See https://www.visitbritain.org/advice-visitors-travel-britain-2019 (accessed 21 June 2019).

Vitic, A. and Ringer, G. (2008) Branding post-conflict destinations: Re-creating Montenegro after the disintegration of Yugoslavia. *Journal of Travel and Tourism Marketing* 23 (2–4), 127–137.

Werbner, P. (2017) Barefoot in Britain – yet again: On multiple identities, intersection(ality) and marginality. *The Sociological Review Monographs* 65 (1), 4–12.

Willett, J., Tidy, R., Trediga, G. and Passmore, P. (2019) Why did Cornwall vote for Brexit: Assessing the implications for EU structural funding programmes. *Environment and Planning C: Politics and Space* 37 (8), 1343–1360.

Wincott, D., Henderson, A., Jeffrey, C. and Wyn Jones, R. (2019) England. In A. Menon (ed.) *Article 50 Two Years On* (pp. 44–45). London: The UK in a Changing Europe. See https://ukandeu.ac.uk/wp-content/uploads/2019/03/Article-50-two-years-on.pdf (accessed 27 June 2019).

4 Uses of the Past: Heritage, Tourism and the Challenges of (Re)Producing Contemporary National Identities in England

Vivian B. Gornik

Introduction

Heritage, the 'present-centered' use of the past (Ashworth, 2007) influences the identities of contemporary citizens (Palmer, 2005; Sommer, 2009). This link between heritage and national identity has been thoroughly established through concepts like the 'invention of tradition' (Hobsbawm & Ranger, 1983) and 'imagined communities' (Anderson, 1991), and scholars have continued to observe and document the homogenising tendency of museums and heritage sites (Drengwitz *et al.*, 2014). Despite this tendency, 'equity, fairness and social justice are of increasing interest among heritage studies and tourism scholars' who argue that heritage tourism has the potential to be a source of social justice (Benjamin & Alderman, 2018: 1).

In this chapter, I explore this relationship between heritage sites and national identity in a changing UK, questioning whether the heritage industry, initially conceived in an era of nation-building, can feasibly play a role in a dynamic reality where identity boundaries are increasingly blurred. Based on 2017 ethnographic fieldwork at Tintagel Castle, Cornwall, south-west England, my discussion illustrates the challenges that heritage tourism sites face in their ability to counteract the historically homogenising tendency of heritage work. I then relate these challenges to the broader issues of British and English identity that have been illuminated by the Brexit referendum and its complex aftermath.

Identity in the United Kingdom

The UK is experiencing an identity crisis. Perhaps it has been since the fall of the Empire, but most recently it has been brought into focus by the results of the referendum on 23 June 2016. On that day, among those who participated, 52% of the British electorate voted to leave the European Union (EU) – the so-called 'Brexit' vote. The results were widely reported as shocking (BBC News, 2016; Bentham & McDonald-Gibson, 2016; Bhatti, 2016; NPR, 2016). Since then, scholars across academic fields have worked to understand this political development (see Alabrese *et al.* [2019] for a recent comprehensive review).

One perspective explains Brexit as a Eurosceptic backlash against the Labour Party's multicultural platform – which speaks directly to concerns about British identity (Clarke & Whittaker, 2016; Hobolt, 2016). The critique of multiculturalism intensified after the 7 July 2005 attacks in London, perpetrated by four British citizens in the name of Islam. Many critics questioned why the perpetrators' national identity did not outweigh allegiance to extremist Islam, while also asking what could prevent others from committing similar crimes (Asari, 2008; Jivraj & Byrne, 2015).

In early 2017, while I was completing my fieldwork, three more terror attacks occurred: (1) on 22 March 2017, a vehicle was weaponised against pedestrians on Westminster Bridge and outside the Palace of Westminster, which houses Parliament; (2) on 22 May 2017, a suicide bomber detonated an explosive in the foyer of the Manchester Arena, as concert-goers were leaving; (3) on 3 June 2017, yet another vehicle was driven into pedestrians, this time on London Bridge. After abandoning the car, the perpetrators then ran into nearby Borough Market where they stabbed several individuals. Although not all of these offenders were British citizens, the 2017 events again raised issues around multiculturalism, immigration and national identity.

Many UK Conservatives have long feared that multiculturalism can breed dangerous segregation, and should be countered, as then-Conservative Party leader David Cameron argued in the wake of the 2005 attacks:

> if we are to bring our society together, then schools – all schools – must teach children that wherever they are from, if they are British citizens, they are inheritors of the British birthright… and every child in our country, wherever they come from must know and deeply understand what it means to be British. The components of our identity – our institutions, our language and our history. (British Political Speech, 2006)

Cameron's 2006 concerns were part of a sustained period of anxiety about the nature of British identity that culminated in the Brexit vote of 2016. Of course, Brexit itself is only the latest specifically political development to raise questions about national identity. Similar questions have surrounded 'devolution' – the transfer of centralised power from the UK Parliament in London to assemblies in Cardiff in Wales, Belfast in Northern Ireland and Edinburgh in Scotland, through legislation in 1997 for Wales and Scotland and 1998 for Northern Ireland.

Further, the 2014 Scottish independence referendum highlighted the question of Scottish identity. A remarkable 84.59% of voters turned out, but the referendum was defeated, with 55.30% voting to remain in the UK. Scotland's strong sense of Celtic identity is shared with Wales and Northern Ireland, and geographic regions like Brittany in France, and the county of Cornwall, where my fieldwork was conducted. Like Welsh, Breton, Irish, Scottish Gaelic and Manx Gaelic, Cornish is a Celtic language. This cultural link between Cornwall and other Celtic regions is part of an established dissonance between Cornishness and Englishness, which I explore in the research.

The case of Cornwall

For centuries, the relationship between Cornwall and the rest of England has been contentious. A centralised London government exploited Cornwall's natural resources in much the same way as in overseas colonies. The Cornish language was suppressed; for example, the 1549 Book of Common Prayer was written only in English as an extension of the English Reformation. The Cornish, along with those from Devon, revolted in what became known as the Prayer Book Rebellion. Although this rebellion was not isolated to Cornwall, at least 2000 Cornish died for the cause and the uprising is an example of the historical tensions between Cornwall and other parts of the country.

In 2014, the Cornish people were recognised by the European Framework Convention for the Protection of National Minorities (Willett, 2014) and, as in Scotland and Wales, a Cornish nationalist movement argues for a devolved legislative Cornish assembly (Woodcock, 2015). This tense relationship is not typically outwardly hostile, but studies show that many Cornish identify as Cornish first, then British, and most likely *not* as English (Woodcock, 2015).

Regionalism throughout England, especially in Cornwall, makes it more difficult to (re)produce homogeneous national identities through national heritage sites without facing some pushback or contestation. Heritage sites in Cornwall find themselves mixed up in the Cornish vs. English tension, especially when heritage sites situated in Cornwall are managed by organizations with names like *English* Heritage.

Today, scholars have highlighted the importance of studying contemporary national and regional identity throughout the UK to better understand the Brexit referendum results (Henderson *et al.*, 2017; Los *et al.*, 2017; Springford *et al.*, 2016). Factors like income inequality, age, gender and attitudes towards immigration have all been compared with voting results across Britain. Henderson *et al.* argue though that:

> Brexit was made in England because of England's population weight in the United Kingdom. And England's choice for Brexit was driven disproportionately by those prioritising English national identity. (Henderson *et al.*, 2017: 643)

This case study then, serves to support the idea that identity is not only complicated when examining the UK as a whole, but it is similarly difficult to pin down within England itself. By examining how national identity is (re)produced at Tintagel Castle, I offer an illustrative insight into the challenges that heritage sites face in addressing issues of identity.

Tintagel Castle and English Heritage

The south-west region of England is known for the beautiful landscapes that characterise many imaginings of the English countryside. The region includes Dartmoor National Park, Exmoor National Park and several places designated as Areas of Outstanding Natural Beauty (AONB) including: the Quantock Hills, Isles of Scilly, Tamar Valley, Blackdown Hills, east Devon, Roseland Heritage Coast, the Cotswolds and the Mendip Hills. The region also boasts two United Nations Educational, Scientific and Cultural Organisation (UNESCO) World Heritage Sites: Stonehenge in Wiltshire and the Jurassic Coast that stretches from Devon to Dorset. In short, south-west England is a magnet for domestic and international tourism.

The research presented here is part of a larger doctoral dissertation project that compared and contrasted two sites linked by Arthurian legend: Tintagel in Cornwall and Glastonbury in Somerset that rely heavily on tourism for their economic survival. Data collection methods included participant observation, semi-structured interviews and archival research. Fieldwork was carried out at Tintagel over a four-week period. A typical fieldwork day included a visit to Tintagel Castle and other parts of the 'heritage complex' such as the Fore Street shops, restaurants and car parks (Robb, 1998). In addition to observing tourists themselves, my interviews focused on a convenience sample of nine heritage 'producers' including individuals working at Tintagel Castle, other local heritage sites and/or nearby tourist amenities. All interviewees quoted here are given pseudonyms to protect their identity.

Tintagel Castle is a massive, multilevel, 18-acre site (Figure 4.1). Through repeated visits on different days of the week, different times of the day and under varying weather conditions, I learned how visitors interact with the site. In town, my observations were similar. Where were

Figure 4.1 View looking out across the Tintagel Castle site (Photo credit: Vivian Gornik)

people parking? Asking for information? Eating? Shopping? I also visited the Royal Cornwall Museum in Truro to conduct archival research, which involved reading old tourist guides of Cornwall. All these factors were relevant in painting a cohesive picture of heritage tourism in Tintagel.

The ruins of Tintagel Castle sit atop a headland that stretches out from the coastline into the ocean. Archaeological evidence reveals it as the location of a large Dark Ages settlement that engaged in significant trading with places as far away as the Mediterranean, and suggests it was home to an important and powerful historical figure. Whether that figure's name was Arthur is not proven. Geoffrey of Monmouth established the Tintagel connection to King Arthur in the 12th century in his *History of the Kings of Britain*, in which he described how Arthur was conceived and born at Tintagel Castle. The impact of this connection is deep and long-standing.

Tintagel Castle has been a notable tourist destination since the mid-19th century. The establishment of the railroad network across Britain in the Victorian era (roughly 1837–1901) made travel easier, which was important for more remote areas like Cornwall, although leisure and tourism remained upper-class activities. In addition, one of the cultural features of the era was a renewed interest in legends like that of King Arthur. Monmouth's influence had made the area famous for centuries, but the decision to change the name of the village from Trevena to Tintagel in the 1850s serves as a marker for the rising importance of tourism to the village.

In 1983, a non-departmental public body of the British Government titled the Historic Buildings and Monuments Commission for England, more commonly referred to as English Heritage, became responsible for the protection and management of a collection of historic properties, including Stonehenge, Dover Castle, parts of Hadrian's Wall and Tintagel Castle. Today, English Heritage's branding tagline is 'Step into England's Story'. Their mission statement states, 'Our vision is that people will experience the story of England where it really happened' (English Heritage, 2018). Their website states:

> We seek to be true to the story of the places and artefacts that we look after and present. We don't exaggerate or make things up for entertainment's sake. Instead, through careful research, we separate fact from fiction and bring fascinating truth to light.

However, the website also states:

> We seek to be imaginative in the way that history is brought to life, thinking creatively, using the most effective means, surprising and delighting people. We want each experience to be vivid, alive and unforgettable.

These two statements illustrate how English Heritage treads the line between education and entertainment – a common dilemma for popular heritage sites.

According to VisitBritain (2019), Tintagel Castle received 229,810 visitors in 2016 and 246,039 in 2017. As one study interviewee remarked:

> I was always really surprised that we are the second busiest site for English Heritage. Certainly Stonehenge is in the stratosphere by comparison, but the number of people we have here, the number of people prepared to pay seven pounds, to go in and see this... they are all coming for a variety of reasons ... but I think primarily it's the idea that Arthur might have been here. (Oliver, English Heritage employee)

The story of Arthur has remained a staple of British culture as each retelling recodes the themes for the audience of its time. Ranging from Monmouth's 12th-century stories and Malory's 15th-century *Le Morte d'Arthur*, to Twain's *A Connecticut Yankee in King Arthur's Court* (1889), White's *Once and Future King* (1958) and even *Monty Python and the Holy Grail* (1975), the appeal of Arthur has never faded. As Radulescu wrote in *Newsweek*:

> In this confusing and sometimes frightening world, audiences seek reassurance in the models of the past. They want a standard of moral

integrity and visionary leadership that is inspirational and transformational in equal measure. One that they cannot find in the world around them, but will discover in the stories of King Arthur. (Radulescu, 2017: 1)

With connections to such a staple of British culture, Tintagel Castle is a unique space in which to examine how heritage sites (re)produce national identity.

Heritage, Identity and Contestation

Today's scholars recognise that viewing heritage as a static thing, passed intact from one generation to the next is inadequate. In *The Uses of Heritage*, Smith (2006: 11) argues that, 'heritage is... ultimately a cultural practice, involved in the construction and regulation of a range of values and understandings'. Heritage is (re)produced in the present and is therefore as much about the present as it is about the past. Ashworth (2007: 35) states, 'heritage is that part of the past that we select in the present for contemporary purposes, whether these be economic or cultural'. Nation-building has been among these purposes.

Nations are 'imagined communities', fused by a cohesive national identity that relies on certain social practices, such as language, ethnicity, religion and other forms of heritage, for a sense of unity (Anderson, 1991). Hobsbawm and Ranger were among the first to claim that heritage was an 'invention' that evolved alongside the nation-state concept. The rise of nation-states in the 19th century required hegemonic, ideologically driven heritage narratives that could unify and ensure social cohesion.

Heritage institutions have continued to reinforce national agendas through 'authorized heritage discourses' (Smith, 2006) and 'heritage regimes' (Geismar, 2015). These discourses can be verbal or non-verbal, expressed through objects and symbols. They do not, however, exist in a cultural vacuum – they are part of how culture is negotiated and contested. So, while heritage discourses may exert hegemonic power, hegemony is never complete; there is always room for counter-hegemonic movements (Franquesa, 2013). The existence of an authorised discourse assumes the potential existence of an unauthorised or counter discourse.

Meskell (2012: 1) states that, 'all heritage work essentially starts from the premise that the past is contested, conflictual and multiply constituted'. That initial homogenising purpose of heritage work requires the erasure of diversity in experience. In *Silencing the Past: Power and the Production of History*, Trouillot (1995: 27) challenges academics involved in studying the past to consider how any historical narrative is 'a particular bundle of silences'. Hall (1999: 67) argues that the production of ideological narratives requires 'marginalising, downgrading or de-legitimating alternative constructions'. As culture changes and silences become more glaring, heritage can become contested. And, as

Bruner (2005: 11–12) describes, contestation can vary on a spectrum from hard to soft, where hard contestation is encountered in spaces where real physical and emotional struggles accompany the ideological and intellectual struggle. There are 'power relations underlying the discourse' which identify 'those who have the ability or authority to speak about or for heritage… and those who do not' (Smith, 2006: 12). Heritage discourses are reflections of the agendas of the powerful; indeed, 'to vilify heritage as biased is thus futile: bias is the main point of heritage. Prejudiced pride in the past is not a sorry consequence of heritage; it is its essential purpose' (Lowenthal, 1996: 122).

Thus, scholars generally agree that heritage was invented to create homogeneous identities – not to illuminate or celebrate diversity. How then, can heritage *do* anything else? Can heritage sites accommodate new calls to engage with social justice (Barton & Leonard, 2010; Bodo, 2012; Fleming, 2016; Jackson, 2011; Nightingale & Sandell, 2012; Smith, 2017; Tribe, 2008)? And does the commodification of heritage for tourist consumption present a significant limitation in doing so?

Selling Heritage

Heritage can become part of the market in the sense that tourists – both domestic and international – have a plethora of options in the types of heritage to consume. Visitors have 'the means to pay for entertainment and heritage sites compete for visitors, thus trying to offer value to the experiences that they are marketing' (Baillie *et al.*, 2010: 55). Kirshenblatt-Gimblett (2006: 193–195) identified a cycle in which processes of valorisation (heritage listing) are followed by those of valuation (working out the income from increased tourism) which can lead to further valorisation, further valuation and so on.

The commodification of heritage inevitably raises concerns about authenticity (Bruner, 2005) as well as cultural loss and the 'collapse of cultural meanings' (Greenwood, 1989). In *Tourism and Legends: Archaeology of Heritage*, Robb (1998: 580) explains his idea of 'heritage complexes', which include 'the commercial exploitation of local heritage themes adjacent to the bounds of official protection'. In his view, the heritage sites that may be demarcated by formal boundaries do not constitute the whole of a consumer's experience of heritage. Rather, the local businesses within the 'complexes exploit heritage resources for profit and help form, reflect, and reinforce visitors' expectations' (Robb, 1998: 580). The idea that heritage tourism is profitable has permeated most parts of the globe, offering the enticing prospect of capitalising on a community's resources. However, the exploitation of heritage resources to meet visitors' expectations may result in accusations of 'Disneyfication'.

Holtorf (2005) argues that rather than fight against the tendency to commodify heritage for entertainment, we should embrace it. Baillie *et al.* (2010: 56) caution that this must be done with an awareness of not 'distorting or trivialising history', presenting a difficult balancing act for both communities and heritage management institutions.

That said, the processes of commodification are becoming democratised in the sense that 'local non-experts can push for or against the development of commodified past' (Baillie *et al*., 2010: 61). For example, King Arthur's Great Halls are a fixture on the Tintagel Fore Street and take full advantage of the Arthur connection. According to George, the site's main manager, many of the bus tours that include Tintagel on their itinerary will also include the Great Halls. Visitors begin by sitting through a 10-minute theatrical 'light show' narration of the story of Arthur before entering the actual Great Halls where 72 stained glass windows depict the life of Arthur and cast multicoloured light on a large Round Table and stone throne. George commented:

> It's an awful thing to say but some people believe. Believe that everything that I've got is real, and obviously it's not. Because there is no proof on King Arthur… but nobody can disprove it. That's the wonderful thing about the legend.

> King Arthur draws them here because of Camelot, the ruins of Camelot down there… King Arthur brings them here from all over the world, and I mean all over the world… South Korea, Vietnam, India, last week um… Brazil, Russia… literally, it amazes me working here, where they come from. Literally everywhere… therefore it benefits the hotels, the guesthouses, the cafes, the shops… some people will come to visit us, purely us and nothin' else, but they usually come here before they go to the ruins or after the ruins.

Baille *et al*. (2010: 69) encourage us to move beyond 'the reification of heritage as priceless and the myth that the commodification of heritage can only be seen as a necessary evil'. Rather, they argue, we should begin investigating commodification as a potentially empowering force for heritage producers and consumers (Baille *et al*., 2010: 69). This project was a small step towards understanding the heritage producers in Tintagel. The contestation of heritage is not just over which narratives are presented and which are silenced, but also the way those narratives might become commodities tied up in endless cycles of 'valorisation and valuation', thereby severely limiting their ability to change. Commodification through heritage tourism can fossilise sites, limiting their ability to account for dynamic cultural realities. So what, if any, identities are at play at Tintagel Castle?

'King Arthur's Castle'

A visitor walking down Fore Street towards Tintagel Castle would come across a variety of establishments ranging from small cafes, a chip shop, a sweet shop and shops with names like Cosmic Gifts, Pendragon Gifts and Little Gems. The local pubs have names like King Arthur's Arms Free House and Guinevere's Restaurant. Lodging options include Avalon Bed and Breakfast and Castle View Bed and Breakfast. These Arthurian inspired names are not a recent development. A 1957 *British Railways Holiday Guide to Southwest England* (1957: 319) includes an ad for 'Camelot Flats Tintagel' described as 'beautifully situated in our grounds overlooking King Arthur's Castle and the Coast of Cornwall'. However, others took care to emphasise the contested nature of the story, such as a 1930 seventh edition of *The Little Guides: Cornwall*:

> In speaking of Arthur, it needs the greatest caution to separate the true, or rather the probable, from the false or improbable. There is absolutely no certainty, except a very certain voice of persistent tradition. (Salmon, 1930: 237)

This illustrates how soft contestation can take place at a site where knowable truth encounters deeply held 'persistent' beliefs and traditions – not unlike those espoused by David Cameron in his speech about British citizenship.

As part of a 2016 rebranding of the Tintagel Castle site, English Heritage installed a new visitor centre exhibition, outdoor interpretive panels and an artistic sculpture. An introductory panel reads:

> Stories of King Arthur have been told across Europe for nearly a thousand years. Tintagel has long been known as the place where the mythical king was conceived, and its history is entwined with the legend. But what really happened here?

In later panels, the responsibility for the Arthurian connection is placed squarely on the shoulders of Geoffrey of Monmouth. This effort to separate fact from fiction is undermined, however, by the installation of a metal sculpture named *Gallos* (Figure 4.2). The official position is that this is *not* Arthur, but visitors generally interpret it as the legendary king. I noted several visitors asking friends or family to take their photo 'with Arthur'. In an interview, even the sculpture artist commented:

> Throughout the whole build process, while I was modelling the piece, it was King Arthur to me. But when it came to near completion, I think [English Heritage] got slightly cold feet because of the whole idea of Disneyfication of the site… they thought that Gallos as a name would suit them.

In a nod to the Cornish heritage of the region, English Heritage chose to name the statue Gallos, Cornish for 'power', to refer to the evidence of

Figure 4.2 Tourists take photos of the sculpture *Gallos* which sits on the upper most level of the Tintagel Castle site (Photo credit: Vivian Gornik)

a powerful Dark Age leader on the site. However, despite the effort to integrate Cornish identity, the power of the larger Arthurian narrative is nearly impenetrable. Arthur has been ideologically fossilised here and is irrevocably linked to Tintagel's heritage landscape. Although there is current demand for the use of this site to be modified, it has been involved in the cycle of valorisation and valuation through the connection to Arthur for too long. Thus, the 'present-centeredness' of this heritage site is significantly limited by tourism.

The inventories of the on-site souvenir shops also reinforce the presence of Arthur. Locally produced beers and ciders, bearing names like Castle Gold, Lancelot Golden Ale and Merlin's Muddle, sit on shelves alongside swords called Avalon Sword and Excalibur. Visitors can buy sword-in-the-stone snow globes of varying sizes, plastic toy dragons, a sorcerer's apprentice hat, a book of spells and replicas of the Gallos statue in the form of figurines, magnets, thimbles, pencil toppers and postcards (Figure 4.3). In fact, Gallos is the most prominent image in the shop. Visitors expect to encounter Arthur at Tintagel, and despite efforts to divert the focus from fiction to fact, the site continues to meet those expectations ten-fold. The village, too, follows suit.

> The village thrives on the castle… they wouldn't have business if it wasn't for the castle… so it's a tricky one really… getting the right balance…

Figure 4.3 Souvenirs inside one of the two English Heritage gift shops at Tintagel Castle (Photo credit: Vivian Gornik)

Even though, you know, there probably wasn't a person here called Arthur, I don't think they're ever going to separate it… it's almost like you've got to have the King Arthur connection in order for people to get interested in the previous history. (Dorothy, English Heritage employee)

Yet, while the dominance of Arthur, a figure of national importance, is undeniable, there are dimensions of this site that are uniquely Cornish.

The 9th edition of *Tintagel Official Souvenir Guide* from the early 20th century describes Tintagel this way:

> Its bleakness and its humanity, its openness and its ancient secrecy… The castle ruins which tradition has associated with King Arthur and his Knights awaken the enthusiasm of sightseers from the world over; but what binds them to Tintagel is the spirit of Cornwall which the place expresses. (Youlton, 1950: 4)

The cliffs and the ocean, the slightly crooked windswept trees, the sound of seagulls and crashing waves, the abandoned tin mines in the distance – this is a Cornish landscape.

Hale (2001: 186) calls Cornwall 'contested territory'. When English Heritage first took over Tintagel Castle and other Cornish sites, protestors would 'argue that English Heritage interpretations are not "authentic": that they do not include or promote education about Cornish history or culture, and that the sites are monarchist, centralist and assimilationist in the narratives they present to the public' (Hale, 2001: 190). One English Heritage employee, Oliver, born and raised in Cornwall, said this:

> The story of north Cornwall, even up until relatively recently, was one of loss. Loss of significant numbers of young people… loss of local councils, that kind of thing… the loss of railways. All of those kind of things… and I think the English have no sense that they are seen as patronisers… they don't ask people what they want, they just come and do it. And there's a regular phrase in Cornwall that we have things done to us.

Oliver's comments highlight the tension between Cornwall and its national government. His decision to say 'the English' draws a distinct line between us, a marginalised Cornish people, and them, the elite, patronising English, who symbolise the London government.

Nearly 20 years after Hale's observations, the tensions remain and the contestation is difficult to resolve, especially at Tintagel Castle. Efforts by English Heritage to incorporate 'Cornishness' into the site are noticeable, particularly the use of the Cornish language on new outdoor panels and on the facades of the ticket office and gift shop. However, the decision to name the statue Gallos instead of Arthur is not, in my opinion, as effective. Visitors see him as Arthur and why should they think otherwise? The omnipresence of Arthurian legend distinctly overshadows any attempts to create a visual representation of a powerful *Cornish* leader. The work of representing Cornish identity has fallen to other smaller, locally managed heritage sites such as St Nectan's Glen, a site of natural beauty renowned for its waterfalls and serene, peaceful atmosphere. The Celtic Clootie tradition of tying strips of cloth, rags or ribbons to trees near sacred wells or springs is alive and well at this site. Although considered 'unspoilt by man', it is not immune to the myth and legend that permeates much of Cornwall's natural and archaeological heritage (St Nectan's Glen, 2019). According to some legends, King Arthur and his Knights were blessed at the main 60-foot waterfall called St Nectan's Kieve.

Conclusion

The tension between regional and national identity is likely to remain in Tintagel as long as the legend of Arthur is fossilised there. The dynamics of the relationship between Cornwall and Westminster are mirrored in other discourses about identity (both regional and national) and voting behaviour in the referendum (Alabrese et al., 2019; Crescenzi et al., 2017; Mckenzie, 2017; Scuira, 2017; Springford et al., 2016; Zoega, 2016). Despite Cornwall's general anti-English sentiment, the county largely voted to leave the EU (Aisch, 2016: 1). Following Henderson et al.'s (2017: 643) argument that 'England's choice for Brexit was driven disproportionately by those prioritising English national identity', Cornwall's vote was, ironically perhaps, indirectly pro-English, even though few might have expressed it in those terms. Based on their findings, Henderson et al. also suggest that:

> [T]he political significance of national identifications in England is an evolving phenomenon, one not closed off by the decision for Brexit. English nationalism, in particular, may play in other ways into the future of UK domestic politics. (Henderson et al., 2017: 644)

For this reason, understanding English identity and the various ways it is produced is of growing importance.

Heritage tourism is a unique arena in which to study issues of identity because, as Sammells (2017: 1) states, 'tourists' gaze can be used to solidify the very national and ethnic boundaries they transgress'. In this study, I have focused on the dynamics of Cornish identity and English identity, as represented by a largely fossilised, legendary narrative. Yet, there are many, more invisible narratives at play in Tintagel and elsewhere. Many of the English who visit Tintagel owe their origins to the UK's colonial past – how can sites like this speak to them? As Hutton (2013: 400) writes, 'a national story intended to generate a common national identity in a period of diversity and change carries the considerable danger of alienating those members of society who cannot instinctually identify with it'. David Cameron's 2006 argument suggests that the answer is forced assimilation, with other cultures learning 'what it means to be British'.

However, the alternative to closing down definitions of English or Britishness, is to expand these definitions – and heritage sites might play a role in this. Certainly, if a site like Tintagel has difficulty in incorporating local Cornish identity into its narrative, it is likely to be even harder to present a more broadly multicultural story. However, there are potential routes; for instance, existing archaeological evidence illustrates how sites like the Tintagel headland were not historically isolated, but connected to places through extensive trade networks. Additional interpretation might challenge the isolationist sentiments of Brexit and foster a more diverse understanding of the site's past. Hutton (2013: 400) suggests that rather than attempting to create coherent but exclusionary narratives about place, 'it may be wiser to emphasise that the land and its heritage are held in common, for people to understand in different ways, but with equal affection and mutual tolerance'. How this noble sentiment may be expressed in practice is the challenge – how to approach broadening the narrative, while continuing to meet the expectations of the tourists upon which so many heritage sites are dependent.

In conclusion, I argue for the salience of studying the uses and expressions of heritage as an entry point into discussions of identity in a rapidly changing UK. These expressions of heritage are always complicated by the pressures that tourism places on the role of heritage in a given community. At some sites, Tintagel arguably among them, visitor expectations have contributed to a rigid narrative, created long ago, within which all future interpretations of the site are likely to reside. But in this mobile, global world there are some heritage sites that have the potential to reframe their narratives of identity in ways that might help promote equity, fairness and social justice.

Acknowledgements

The research presented herein was funded in part by two sources: (1) a visiting researcher scholarship from the College of Humanities at the University of Exeter and (2) travel funding from the Humanities Institute at the University of South Florida. Thank you also to Dr Elizabeth Bird (University of South Florida) for serving as my mentor throughout the fieldwork and writing process associated with the dissertation research examined here.

References

Aisch, G., Pearce, A. and Russell K. (2016) How Britain voted in the E.U. referendum. *The New York Times*. See https://www.nytimes.com/interactive/2016/06/24/world/europe/how-britain-voted-brexit-referendum.html (accessed 18 July 2019).

Alabrese, E., Becker, S., Fetzer, T. and Novy, D. (2019) Who votes for Brexit? Individual and regional data combined. *European Journal for Political Economy* 56, 132–150.

Anderson, B. (1991) *Imagined Communities: Reflections on the Origin and Spread of Nationalism*. London/New York: Verso.

Asari, E. (2008) British national identity and the dilemmas of multiculturalism. *Nationalism and Ethnic Politics* 14 (1), 1–28.

Ashworth, G.J. (2007) *Pluralising Pasts: Heritage, Identity and Place in Multicultural Societies*. London/Ann Arbor, MI: Pluto Press.

Baillie, B., Chatzoglou, A. and Taha, S. (2010) Packaging the past: The commodification of heritage. *Heritage and Society* 3 (1), 51–71.

Barton, A. and Leonard, S. (2010) Incorporating social justice in tourism planning: Racial reconciliation and sustainable community development in the Deep South. *Community Development* 41 (3), 298–322.

BBC News (2016) 'EU Referendum: The results in maps and charts'. See https://www.bbc.com/news/uk-politics-36616028 (accessed 18 July 2019).

Benjamin, S. and Alderman, D. (2018) Performing a different narrative: Museum theater and the memory-work of producing and managing slavery heritage at southern plantation museums. *International Journal of Heritage Studies* 24 (3), 1–13.

Bentham, M. and McDonald-Gibson, C. (2016) EU Referendum Result Shocks World Leaders as Britain Backs Brexit. *Evening Standard*, 24 June. See https://www.standard.co.uk/news/politics/damn-eu-referendum-result-shocks-world-leaders-as-britain-backs-brexit-a3280031.html (accessed 1 July 2017).

Bhatti, J. (2016) Europeans Shocked, Worried Over Brexit. *USA Today*, 24 June. See https://www.usatoday.com/story/news/world/2016/06/24/europeans-brexit-european-union-britain/86326188/ (accessed 1 July 2017).

British Political Speech (2006) Leader's speech, Bournemouth, David Cameron (Conservative), 4 October 2006. See http://www.britishpoliticalspeech.org/speech-archive.htm?speech=151 (accessed 21 October 2016).

British Railways Holiday Guide to Southwest England (1957) Guidebook. Courtney Library at Royal Cornwall Museum (accessed 21 June 2017).

Bodo, S. (2012) Museums as intercultural spaces. In E. Nightingale and R. Sandell (eds) *Museums, Equality and Social Justice* (pp. 181–191). London: Routledge.

Bruner, E. (2005) *Culture on Tour: Ethnographies of Travel*. Chicago, IL: University of Chicago Press.

Clarke, S. and Whittaker, M. (2016) The importance of place: Explaining the characteristics underpinning the Brexit vote across different parts of the UK. *Resolution Foundation*, 15 July. See https://www.resolutionfoundation.org/publications/the-important-of-place-explaining-the-characteristics-underpinning-the-brexit-vote-across-different-parts-of-the-uk/ (accessed 1 July 2019).

Crescenzi, R., Cataldo, M. and Faggian, A. (2017) Internationalized at work and localistic at home: The 'split' Europeanization behind Brexit. *Papers in Regional Science* 97, 117–132.

Drengwitz, B., Elbers, B., Jahn, L. and Wrogemann, I. (2014) Nation and national museums, a contested relationship: An analysis of U.S. national museums in the twenty-first century. *Curator: The Museum Journal* 57 (1), 97–106.

English Heritage. Our Vision and Values. See http://www.english-heritage.org.uk/about-us/our-values/ (accessed 15 May 2018).

Fleming, D. (2016) Do museums change lives?: Ninth Stephen Weil Memorial Lecture. *Curator: The Museum Journal* 59 (2), 73–79.

Franquesa, J. (2013) On keeping and selling: The political economy of heritage making in contemporary Spain. *Current Anthropology* 54 (3), 346–369.

Geismar, H. (2015) Anthropology and heritage regimes. *Annual Review of Anthropology* 44, 71–85.

Greenwood, D. (1989) Culture by the pound: An anthropological perspective on tourism as cultural commoditization. In V. Smith (ed.) *Hosts and Guests: The Anthropology of Tourism* (pp. 171–185). Philadelphia, PA: University of Pennsylvania Press.

Hale, A. (2001) Representing the Cornish: Contesting heritage interpretation in Cornwall. *Tourist Studies* 1 (2), 185–196.

Hall, S. (1999) Un-Settling 'The Heritage', Re-Imagining the Post-Nation. Paper presented at Whose Heritage? The Impact of Cultural Diversity on Britain's Living Heritage. G-Mex, Manchester, England.

Henderson, A., Jeffrey, C., Wincott, D. and Jones, R. (2017) How Brexit was made in England. *The British Journal of Politics and International Relations* 19 (4), 631–646.

Hobolt, S. (2016) The Brexit vote: A divided nation, a divided continent. *Journal of European Public Policy* 23, 1259–1277.

Hobsbawm, E. and Ranger, T. (1983) *The Invention of Tradition*. New York: Cambridge University Press.
Holtorf, C. (2005) Beyond crusades: How (not) to engage with alternative archaeologies. *World Archaeology* 37 (4), 544–551. doi: 10.1080/00438240500395813.
Hutton, R. (2013) *Pagan Britain*. New Haven, CT: Yale University Press.
Jackson, A. (2011) Shattering slave life portrayals: Uncovering subjugated knowledge in U.S. plantation sites in South Carolina and Florida. *American Anthropologist* 113 (3), 448–462.
Jivraj, S. and Byrne, B. (2015) Who feels British? In S. Jivraj and L. Simpson (eds) *Ethnic Identity and Inequalities in Britain: The Dynamics of Diversity* (pp. 65–79). Bristol: Policy Press.
Kirshenblatt-Gimblett, B. (2006) World heritage and cultural economics In I. Karp, C. Kratz, L. Szwaja and A. Ybarra-Frausto (eds) *Museum Frictions. Public Cultures/ Global Transformations* (pp. 161–202). Durham, NC: Duke University Press.
Los, B., McCann, P., Springford, J. and Thissen, M. (2017) The mismatch between local voting and the local economic consequences of Brexit. *Regional Studies* 51 (5), 786-799.
Lowenthal, D. (1996) *Possessed by the Past: The Heritage Crusade and the Spoils of History*. New York: The Free Press.
Mckenzie, L. (2017) The class politics of prejudice: Brexit and the land of no-hope and glory. *The British Journal of Sociology* 68 (1), 265–280.
Meskell, L. (2012) *The Nature of Heritage: The New South Africa*. Walden, MA: Wiley-Blackwell.
Monmouth, G. (1136) *The History of the Kings of Britain*.
Nightingale, E. and Sandell, R. (2012) Introduction. In E. Nightingale and R. Sandell (eds) *Museums, Equality and Social Justice* (pp. 1–9). London: Routledge.
NPR (2016) Shocked by Trump win, Brits see Brexit parallels and commiserate. *NPR Morning Edition*, 9 November. See https://www.npr.org/sections/parallels/2016 /11/09/501370571/shocked-by-trump-win-brits-see-brexit-parallels-and-commiserate (accessed 1 December 2016).
Palmer, C. (2005) An ethnography of Englishness: Experiencing identity through tourism. *Annals of Tourism Research* 32 (1), 7–27.
Radulescu, R. (2017) Why the legend of King Arthur still resounds today. *Newsweek*, 2 February. See https://www.newsweek.com/king-arthur-round-table-myth-literature -552155 (accessed 18 July 2019).
Robb, J. (1998) Tourism and legends: Archaeology of heritage. *Annals of Tourism Research* 25 (3), 579–596.
Salmon, A. (1930) *The Little Guides: Cornwall*. London: Methuen and Co.
Sammells, C. (2017) Session Abstract: 'Tourism and Mobility Matters'. American Anthropological Association Meeting, Washington, DC, 2 December.
St Nectan's Glen (2019) Saint Nectan's Glen history. See https://www.st-nectansglen.co.uk /history/ (accessed 2 September 2019).
Scuira, L. (2017) Brexit beyond borders: Beginning of the EU collapse and return to nationalism. *Journal of International Affairs* 70 (2), 109–123.
Smith, L. (2006) *Uses of Heritage*. New York: Routledge.
Smith, L. (2017) 'We are… we are everything': The politics of recognition and misrecognition at immigration museums. *Museums and Society* 15 (1), 69–86.
Sommer, U. (2009) Methods used to investigate the use of the past in the formation of regional identities. In M.L. Sorensen and J. Carman (eds) *Heritage Studies: Methods and Approaches, Volume 1* (pp. 103–120). New York: Routledge.
Springford, J., McCann, P., Los, B. and Thissen, M. (2016) Brexiting yourself in the foot: Why Britain's Eurosceptic regions have the most to lose from EU withdrawal. *Centre for European Reform*. See https://www.cer.eu/insights/brexiting-yourself-foot-why -britains-eurosceptic-regions-have-most-lose-eu-withdrawal (accessed 18 July 2019).

Tribe, J. (2008) Tourism: A critical business. *Journal of Travel Research* 46 (3), 245–255.

Trouillot, M. (1995) *Silencing the Past: Power and the Production of History*. Boston, MA: Beacon Press.

VisitBritain (2019) Annual Survey of Visits to Visitor Attractions. See https://www.visitbritain.org/annual-survey-visits-visitor-attractions-latest-results (accessed 18 July 2019).

Willett, J. (2014) Cornwall's new status shows how regionalism is changing nations everywhere. *The Conversation*, 24 April. See https://theconversation.com/cornwalls-new-status-shows-how-regionalism-is-changing-nation-states-everywhere-25926 (accessed 1 December 2016).

Woodcock, P. (2015) Cornwall and Yorkshire show regional identities run deep in England, too. *The Conversation*, 12 May. See https://theconversation.com/cornwall-and-yorkshire-show-regional-identities-run-deep-in-england-too-41322 (accessed 1 December 2016).

Youlton, R.J. (1950) *Tintagel and Boscastle, North Cornwall: Official Souvenir Guide*. Tintagel: Fort Knox Museum.

Zoega, G. (2016) On the causes of Brexit: Regional differences in economic prosperity and voting behaviour. *VoxEU*, 1 September. See https://voxeu.org/article/brexit-economic-prosperity-and-voting-behaviourn (accessed 19 July 2019).

5 Royal Events and Tourism in the Post-Brexit Era

Jennifer Frost and Warwick Frost

Introduction

In 1970, during a period when Britain debated whether to join the European Economic Community, the first official walkabout of a member of the British royal family occurred outside Wellington Town Hall in New Zealand. The use of the term is attributed to Vincent Mulchrone of the *Daily Mail*, in a headline that read 'Queen goes walkabout' (McIntyre, 1991: 250). Rather than merely seeing members of the royal family pass by in a procession or motorcade, waving at the crowds, it allowed members of the public to directly meet and speak with royalty, even if behind a barricade or rope. The idea was attributed to Sir Patrick O'Dea, the New Zealand Secretary for Internal Affairs (Hardman, 2012), and promised to shine a spotlight on a high-profile occasion that the New Zealand government was keen to celebrate – the Bicentenary of Captain James Cook's voyage to that country. This walkabout literally broke down some of the barriers between sovereign and subject (Bramston, 2015). On this first occasion, the Queen spoke with those waiting to see her for roughly half an hour (McIntyre, 1991), and its success meant that it was repeated in Australia and then back in England in Coventry (Harrison, 1996). The walkabout subsequently became one of the rituals we now expect to see when the British royal family attend public engagements and is an example of an invented tradition (Cannadine, 1983; Hobsbawm, 1983), as well as highlighting the way that royal events such as an official tour can be subtly harnessed for political purposes.

Apart from improving access, the introduction of the walkabout also increased interest in the royal family; at a time when there was concern about their relevance and popularity. This was just a year after the 'warts and all' film *Royal Family* (1969) was screened on television, showing Prince Philip barbecuing meat while on holiday at Balmoral and the Queen eating her breakfast (Laing & Frost, 2018). It also occurred during a period when Great Britain became interested in joining the Common Market with Europe; an application which was initially refused in

1967, in part because of France's President de Gaulle's distrust of Britain, calling it 'an American "Trojan horse"' (Chace, 1973: 98). Britain subsequently signed the treaty to join the Common Market in 1973. At this time, it was keen to show its bona fides towards Europe, while not alienating members of the Commonwealth and strong trading partners such as Australia and New Zealand. Unveiling the walkabout in New Zealand was a masterstroke, for that country was seen as the most British of the settler societies and potentially had the most to lose in exports of primary products. It illustrated how the royal family could be used strategically to emphasise historical links and relationships and smooth troubled political waters. This role can be seen in the present day, usually connected to royal events such as tours or weddings, and has been particularly prominent in the wake of the Brexit vote in 2016, when a majority of Britons voted to leave the European Union (EU).

There is a long history of royal events functioning to stimulate national pride and allegiance – as we argue in our book *Royal Events: Rituals, Innovations, Meanings* (Laing & Frost, 2018). Typically, this view is inward-looking, focusing on how such events influence domestic sentiment. However, royal events and ceremonies may also send messages to external actors, including potential allies and trade partners. In this chapter, our aim is to examine how royal events and tourism have been subtly refashioned in order to accommodate and strengthen new international relationships in the post-Brexit era.

Royal Events as a Political Tool

There is a broad range of events that could be characterised as *royal events*, including high-profile ceremonial events such as weddings, funerals and coronations; royal tours; and public engagements, often involving patronages or charities supported by members of the royal family (Laing & Frost, 2018). Their role and function is of interest to scholars given their ubiquity, the costs that underpin them, the revenue that is said to flow from them and their potential social impacts, such as fostering community identity, social capital and inclusion (Laing & Frost, 2018; Long, 2008; Prochaska, 1995). They are also innovative, and influence event management more broadly through the demonstration effect that occurs when organising successful large-scale ceremonial occasions.

One of the most frequently advanced arguments for maintaining the British royal family in grand style, with crowns, carriages and castles, rather than an informal 'bicycle' style monarchy like the Netherlands (Cannon, 1987), is that this grandeur is said to encourage tourism. A 2017 report by Brand Finance, a British-based consultancy that specialises in advice on maximising the value of organisational brands, argued that the royal family is responsible for £550 million of international tourism to Great Britain (Specter, 2017), and it has been observed that these visitors expect to see a royal ceremony and see this as the 'chief function' of royalty (Long, 2008: 3). Certainly, annual extravaganzas such as the Trooping of the Colour (Figure 5.1) and one-off events like the royal wedding of Prince Harry and Meghan Markle in 2018 attract extensive media coverage and large crowds and keep the monarchy visible to the public.

Alternatively, royal events have been criticised for reinforcing elitism and class boundaries; for attracting government money that could have been spent elsewhere; and for being inauthentic, often based on traditions that have been invented, in part to serve political ends (Brunt, 1992; Cannadine, 1983; Hobsbawm, 1983; Nairn, 1988). It is noteworthy that most of this literature dates from the 1980s and 1990s, at a time when the royal family was attracting public criticism over its relevance and a series of scandals, such as the divorce of the Prince and Princess of Wales. Cannadine (1983), for example, refers to the use of carriages in royal processions as anachronistic, originating in a time when this was a standard mode of transport for many people, and being passed off as a ritual long after the time when they had been retired in broader society in favour of more modern technology such as the motor car. Others such as Kuhn (1996) argue that this invention of traditions argument is flawed, in that it fails to recognise that the British monarchy has always reinvented itself, which is part of the reason why it still survives when many of its counterparts internationally have not. There is ultimately a pragmatism at play here, even if it can take some time to manifest itself, whether it be

Figure 5.1 Parade during the 1988 Trooping of the Colour (Source: Margaret Zallar, authors' collection)

the Queen's (reluctant) decision to lower the flag in the wake of Princess Diana's death in 1997, against royal protocol, or the decision by recent royal brides to travel to their weddings in a car rather than a carriage. The monarchical brand has been described as incorporating the five Rs – royal, regal, relevant, responsive and respected (Balmer *et al.*, 2006) and 'the mention of relevance and responsiveness is a direct acknowledgement of the importance of staying in touch in a changing world, and innovating where necessary' (Laing & Frost, 2018: 231).

It is fair to say, whatever one's views are on their value, that royal events are rarely staged without having some *political* implication, regardless of the ideal that the constitutional monarch is to be separated 'from [government] business', which 'preserves its mystery' (Bagehot, 1867). This is not to say that individual members of the royal family have not had their political hobby horses. Prince Charles is the most high-profile example of this, with his outspoken views often seen as contrary to government policy, although in some areas, such as climate change, they might now be seen as ahead of his time (Hardman, 2012). The monarch herself, however, has remained aloof from political issues, and recent comments from the Prince of Wales in the documentary *Prince, Son and Heir: Charles at 70* (2018) have made it clear that he subscribes to the importance of this: 'I'm not that stupid. I do realise that it is a separate exercise being sovereign. So of course I understand entirely how that should operate'. The political dimension of the family has itself changed over time, evolving: 'from an institution with dwindling political power into a tremendously attractive centerpiece of national identity' (Kuhn, 1996: 10). This role, by definition, is a political construct, with governments keen to 'create among the governed a collective and tacit identity' (Dobransky & Fine, 2006: 559), which results in the belief that 'power arrangements are natural and genuine' (Dobransky & Fine, 2006: 559–560). The royal family is one of the tools used by the British government in pursuit of this goal, symbolising the nation and stability during periods of upheaval. There is a symbiotic relationship, in that the monarchy's political relevance arguably protects them from moves towards republicanism.

A salient example of this occurred in 1997, in the wake of the public criticism of the Palace's response to Princess Diana's death. Prime Minister Tony Blair worked to broker a truce between the Palace old guard and increasingly militant members of the media, as well as arguably the public, and by doing so, 'effectively saved the monarchy in Britain at a time when it could have been seriously damaged' (Haseler, 2012: 168). The destruction of the monarchy was not in Blair's interests, while his connection with Diana in the minds of the public, through comments like 'the people's princess', arguably increased his popularity at the time. As Australians, we are conscious that the republican debate has diminished in our country, following an unsuccessful referendum in 1999.

In this chapter, we will look at case studies to illustrate this relationship between politics and the royal family, through the aegis of a royal event. We compare events from the 1950s, such as the 1953 coronation of Queen Elizabeth II and the royal tour of the Commonwealth by the Queen and the Duke of Edinburgh in 1953/1954, with more recent occurrences in the Brexit era. Their common denominator is the centrality of the Commonwealth in these high-profile events, and how this has helped the governments of the day achieve their political ends.

The Coronation of Queen Elizabeth II

At the time of Queen Elizabeth's coronation in 1953, there was a heavy emphasis on her role as head of the Commonwealth, rather than her European roots, which reinforced political objectives of the time. The nation was still recovering from war, and it was seen as important to show that Britain was still a force with which to be reckoned. Maintaining political ties with its new Commonwealth of former imperial colonies was therefore a crucial element of this national narrative. While the coronation 'was, naturally, a nationalistic occasion… it was also one through which Britons attempted to negotiate their relationship to the new world order, an order in which the fabled British Empire was no more, and its replacement, the Commonwealth of Nations, had yet to articulate its role or prove its effectiveness' (Feingold, 2009: 148).

This goal of using the coronation to emphasise Britain's glorious past – and future – dovetailed with the royal family's own desire to be seen as quintessentially British and at the heart of the nation. While Elizabeth was from a family of German ancestry, the name of the royal house of Saxe-Coburg-Gotha had been changed to the House of Windsor in 1917, due to concern of King George V that the name allied the family too closely with their European foes. He wanted a British name to reassure the public that their interests were bound up with their own, and chose the name of the beloved castle that is the oldest occupied royal residence in the world. The king was particularly conscious of the number of their foreign relatives who had either been deposed or assassinated in the lead up to the First World War, and anti-German feeling throughout Britain was on the rise. Just a few months earlier, his cousin, Tsar Nicholas II of Russia, had been forced to abdicate his throne, and the following year, he and his family were killed by firing squad by the Bolsheviks. It was a climate in which George V feared for his own family's survival.

The king was successful in reinventing the family with a strong British identity, especially when his second son Albert, later to become King George VI, married a popular British aristocrat, Lady Elizabeth Bowes-Lyon. The couple had two photogenic children, the Princesses Elizabeth and Margaret Rose, and their wholesome family life, complete with dogs and brisk country walks, captured in women's magazines and newspapers, was in sharp contrast to the hedonistic lifestyle of the Prince of Wales, who later foreswore his throne for his American lover, divorcee Wallis Simpson (Laing & Frost, 2018). Their decision to stay in London during the Blitz was also looked on favourably. History shows that in times of crisis, such as wartime, the propaganda value of an event involving a royal personage can be very high and help to raise morale. Prochaska (1995: 223) refers to the engagements carried out by the then Queen Elizabeth, later the Queen Mother, during the Second World War: 'Day after day, always on the move, [the Queen] consoled the weary and the afflicted. Arguably, the tours during the blitz to the East End and the docks, Swansea and Plymouth, did more to impress on the public the caring and compassionate role of the monarchy than all her patronage work put together' (Prochaska, 1995: 223). The family became a focus for morale in a bleak and frightening time:

> A large proportion of the public felt a connection with them as a family undergoing the same struggles as they did during the war, and who were now looking ahead in the hope of better times. Thus a wedding of their eldest daughter was a moment of national celebration and pride. (Laing & Frost, 2018: 165)

Placing the Commonwealth at the heart of the coronation also coalesced with the new Queen's own political views. She saw it as one of the legacies of her father's short reign (Hardman, 2012). Thus, several symbolic tributes to the Commonwealth were incorporated into the ceremony (Feingold, 2009; Richards, 2004; Royal Family, 2019). Commonwealth leaders and representatives formed part of the Sovereign's procession, while troops from the Commonwealth took part in the procession through the streets. Flowers from across the Commonwealth filled Westminster Abbey, while Queen Elizabeth's gown (Figure 5.2), designed by Norman Hartnell, featured embroidery of eleven floral emblems, one each for the four countries of the United Kingdom, and those representing the seven countries of the Commonwealth. The latter included the Australian wattle, the Indian lotus flower, the New Zealand fern, the South African protea and the Canadian maple leaf. Even the newspaper headlines on the morning of the coronation, announcing the ascent of Everest by the New Zealander Edmund Hillary, and playing down the involvement of the Nepalese Tenzing Norgay, painted the achievement as an example of British imperial might (Haseler, 2012; Laing & Frost, 2018). Televising the event for the first time allowed a global audience to participate in the ritual, rather than a select few (Feingold, 2009). The coronation was therefore seen as an important watershed in relations between the Crown and its subjects, who would not have missed the symbolism, even if they wanted to. Richards (2004: 71) notes that 'The day after the coronation, *The Times* observed that the sense of Empire

Figure 5.2 Promotional poster for the *Tudors to Windsor: British Royal Portraits* Exhibition, Bendigo Art Gallery, Australia, depicting the coronation gown of Queen Elizabeth II (Source: Jennifer Frost)

so great at the time of Queen Victoria's golden jubilee had been replaced by a new feeling of commonwealth community'.

While the Queen's dress might have been seen as innovative, there was actually a long-standing tradition of using royal fashion to make a political statement, particularly in the context of these high-profile royal events. For example, imperial symbolism was incorporated into the clothing worn by the Danish-born Queen Alexandra at the coronation of her husband King Edward VII in 1902. The Queen wore an overdress designed and woven in India; a decision ascribed to her fascination with the country she had never visited. The then Vicereine of India, Lady Curzon, described it in a letter to her father in 1901 as 'embroidered in India with all the emblems of the Empire worked into a fine design' (quoted in Strasdin, 2012: 157). It represented 'Indian exoticism' (Strasdin, 2012: 161), making the Queen look like a vision from a fairy tale, rather than a stiff monarch (Strasdin, 2012). Apart from acting as a graceful homage to India from its Empress, the dress also had the benefit of secrecy, having been made overseas, and thus was less likely to be copied (Strasdin, 2012). Alexandra knew how important it was that she take centre stage as the consort to the king, while also emphasising her – and his – prominence within the Empire. The power of fashion was even more profound in the case of a woman who would rarely be heard by her subjects. Her clothing had to do the talking.

The Royal Tour of the Commonwealth in 1953/1954

The public relations coup of the coronation was followed by a high-profile tour by Elizabeth and her husband Philip. Its purpose was to take the new Queen to the Commonwealth, and to strengthen relations through being seen *in situ* at public events. In her Christmas broadcast of 1953, recorded in Auckland, New Zealand, Elizabeth not only brought up the comparisons that had been made between her new reign and that of her famous forebear, Elizabeth I, but also distinguished *her* realm from Europe. She suggests that Britain's resilience and achievements should not be underestimated: 'For her Kingdom, small though it may have been and poor by comparison with her European neighbours, was yet great in spirit and well endowed with men who were ready to encompass the earth' (Royal Household, 2017). She also emphasised the vital role played by the Commonwealth, 'built on the highest qualities of the spirit of man: friendship, loyalty, and the desire for freedom and peace'. The message delivered to her audience is of 'Britain's centrality in this new assemblage of nations' (Feingold, 2009: 151).

The 1953 tour was mostly spent in Australia and New Zealand, seen as important destinations for 'reciprocal reinforcement of shared imperial (now Commonwealth) identity' (Feingold, 2009: 149). Britain was particularly keen to ensure that its relationship with Australia was not supplanted by the United States, given the closer ties that had developed between those two nations during the Second World War. The decision to have the Queen present at the opening of the American War Memorial in Canberra was therefore a political one; 'an attempt to weaken growing American influence' (Alomes, 1988: 55). The tour was also taken seriously by most politicians from these former dominions, regardless of their political persuasion, who saw mileage in being associated with the glamour of the visit (Spearritt, 1988).

Official engagements during the tour were designed to allow the royal couple to be seen by as many members of the public as possible, and this was facilitated by government initiatives that made it almost impossible not to be aware of the tour or to take part in it. For example, in New Zealand, funding was provided that allowed all schoolchildren to be bussed in to see the Queen (Feingold, 2009). One of these schoolboys was

David Lange, who later became the prime minister of New Zealand, and who remembered that 'roads were sealed so she could drive along them, or in the case of the road from Kaikohe to Whangarei, the half she drove on was sealed and the other half finished many years afterwards' (New Zealand Ministry for Culture and Heritage, 2014). Souvenirs were readily available, and the streets were lined with people waving flags, eager to catch a glimpse of royalty. In Sydney, 'more than a quarter of a million people turned out just to watch the Queen return from the theatre' (Hardman, 2012: 6). Not all the public involvement was passive – there were numerous performances in front of the Queen and the Duke by local people, sometimes with an indigenous theme, but also incorporating British anachronisms such as maypole dancing (Connors, 1993) and dyeing sheep and planting flowers in hues of red, white and blue (New Zealand Ministry for Culture and Heritage, 2014). Criticism has been levelled at the tour for being jingoistic and kitsch, with Alomes (1988: 155) judging it as 'the culmination of a century of imperial indoctrination'. Whatever one's views on the way it was planned and carried out, it cemented the popularity of the royal family and the relationship with Britain at this important post-war juncture (Connors, 1993), and must be judged a political success on this basis.

As occurred at the coronation, clothing was used to reinforce the message of Commonwealth unity during the tour. The Queen's coronation gown, with its symbolic floral motifs, was worn on tour when opening the parliaments of Australia, New Zealand and Ceylon (now Sri Lanka) in 1954 and Canada in 1957, and made a powerful political statement. Wearing the gown in these places and on these occasions gave members of the Commonwealth a sense of the connection between their country's 'common symbolic currency' (Dobransky & Fine, 2006: 561) and the Crown. The Queen also wore day clothing replete with national motifs, such as, for example, a hat in Canberra decorated with wattle (Laing & Frost, 2018), whose symbolic message was clear to all who saw the numerous photographs in newspapers and magazines at the time.

Royal Events in the Post-Brexit Era

At the start of the 21st century, the political imperatives around the relationship between Great Britain and the Commonwealth had changed. It was increasingly seen as outdated – 'an irrelevant imperial hangover' (Hardman, 2012: 300), although the monarch herself still saw it as one of the greatest achievements of her reign. The political focus was arguably the relationship with Europe, aligning with Britain's economic interests, even though there had been persistent disquiet in some quarters about the alleged deleterious effect on British sovereignty and 'the British public has consistently been the most Eurosceptic electorate in the EU ever since the UK joined in 1973' (Hobolt, 2016: 1259–1260). However, in 2011, the

British Foreign Secretary William Hague (2011) argued that the Commonwealth was now 'back at the very heart of British foreign policy for the first time in more than a decade... it is a cornerstone of our foreign policy, alongside our role in the EU, our membership of NATO and our Special Relationship with the United States of America... In a world that is dominated by networks and not by the power blocs of old, the Commonwealth is the ultimate network'.

In 2016, prior to the Brexit vote, the Commonwealth spokesperson for the UK Independence Party went further, arguing 'Outside the EU, the world is our oyster, and the Commonwealth remains that precious pearl within' (Lowe, 2016). This viewpoint was an important philosophical underpinning of the Leave campaign, which 'presented the Commonwealth as a viable economic alternative to the EU, with [the] bonus of working with an organisation that already had historical connections to the UK' (Namusoke, 2016: 468). It also reminded the public of a time when they perceived that Britain had greater autonomy and prestige, which might be argued to have racial overtones and appeal to those who considered themselves an underclass or a forgotten voice in modern Britain.

As soon as the maelstrom that was Brexit occurred, the royal family were galvanised to try to repair some of the damage. The trade relationship with Europe was considered to be still important, even if border controls were to be enforced. Almost immediately after the vote, there were hastily arranged royal visits to European countries by Prince Charles and the Duchess of Cornwall, and Prince William and the Duchess of Cambridge, in order to reassure them that there would still be linkages, albeit outside the European Community. The visit by William and Kate to Paris, said to have been requested by the Foreign Office (Laing & Frost, 2018), was arguably overshadowed by headlines the week before about the Prince's failure to attend the annual Commonwealth Day service, in preference for a ski trip with friends, complete with laddish behaviour and drunken antics. It sent the wrong message about the importance of foreign relations to the Palace, especially given the personal significance that this regular engagement on the royal calendar is said to have for the Queen (Hardman, 2012). However, the French trip was regarded as a success, thanks to photogenic moments of the Duchess wearing French couture (Chanel) in front of the Eiffel Tower, and the sincerity of the couple when meeting survivors of a recent terrorist attack. The UK government were transparent about the perceived benefits of the tours for smoothing over troubled waters post-Brexit: 'The Royal Family are excellent and experienced ambassadors for the United Kingdom. Whilst every royal visit is unique, each visit is designed to support foreign policy objectives and promote closer ties across a range of areas, for example cultural, economic or political, between the UK and the host country' (BBC, 2017). Royal correspondent Peter Hunt summed these tours up as follows: 'Number 10 can portray the trip as a sign of Prime Minister Theresa May's words in action that we may be leaving the European Union, but we want to remain firm friends with Europe' (BBC, 2017).

In the lead-up to the 2018 royal wedding between Prince Harry and Meghan Markle, the global media coverage speculated on what it meant for Britain and its place in the modern world. One key trope was that the wedding would boost national morale, functioning as a tonic to the confusion and pessimism that had dominated the country's psyche since the Brexit vote. It shifted the emphasis beyond Europe in two ways. The first was in strengthening ties with the United States. This was perhaps not surprising given that the bride was American, but it was further reinforced through features of the ceremony and the strength of the American media coverage. The second was a focus on the Commonwealth, and the shared identity of a range of countries as members of that organisation. Coming only a few months after the Commonwealth Games, the wedding also coincided with much discussion about future trade deals returning Commonwealth countries to something like their pre-European Community position. It was announced at the time of the couple's engagement that they would have special roles associated with promoting the Commonwealth, and the bride-to-be spoke of her desire 'to just really get to know more about the different communities here, smaller organisations, we're working on the same causes that I've always been passionate about under this umbrella. And also being able to go around the Commonwealth I think is just the beginning of that'. The Duchess of Sussex was appointed as patron of the Association of Commonwealth Universities and the vice president for the Queen's Commonwealth Trust, and Prince Harry was designated president of the Commonwealth Trust; with these roles continuing after their decision to step down from their roles as senior members of the royal family in 2020, commonly referred to as Megxit.

At her wedding, Meghan followed in the footsteps of the Queen, using symbolic floral emblems on her ceremonial garb. In her case, it was her bridal veil, embroidered with 53 flowers from across the Commonwealth countries. In a recent documentary *Queen of the World*, the Duchess revealed that it was a surprise for her husband on their wedding day, as well as other members of the royal family: '[Harry was] really over the moon to find out that I would make this choice for our day together. I think the other members of the family had a similar reaction'. She noted that there was 'appreciation for the fact that we understand how important this is for us and the role that we play, and the work that we're going to continue to do within the Commonwealth countries'. The wedding also incorporated an American gospel choir and an African-American Episcopal priest, which made headlines for being so different from the planning behind previous royal weddings. There was no attempt to underplay the bride's racial or national origins. The guest list was replete with American celebrities such as George Clooney and Oprah Winfrey, rather than blue-blooded European royalty, while bridal attendants Prince George and Princess Charlotte rubbed shoulders with the children of a popstar and the bride's friends, including a Canadian fashion stylist and an American former entertainment lawyer and entrepreneur. The millions of Americans watching the coverage could not have failed to see the tributes that were paid to their country, and, at the time, the woman who was being welcomed with open arms as a breath of fresh air into the royal family.

Public interest in the wedding was extremely high, tapping into memories of past experiences, such as the post-war wedding of Queen Elizabeth and the wedding of Charles and Diana in 1981. Like them, this was portrayed as an event of global significance. In Australia, for example, it was telecast in prime time on four of the five national television networks. The commercial Channel 7 network even branded itself the 'Royal Channel'. The tourism impacts of the event were strongly emphasised during this coverage, based on nostalgic imagery of Britishness, with the royal family at the heart of that narrative. The overwhelming impression conveyed through the media was that of the importance and enduring qualities of British traditions, allied with the Cinderella myth, with an American divorcee being accepted as a princess.

The first official tour of the Duke and Duchess of Sussex after their marriage was to Australia, New Zealand, Fiji and Tonga, again emphasising the special relationship between Great Britain and the Commonwealth. Several tours since then, however, have demonstrated a willingness to interact with new countries that are neither within the EU nor the Commonwealth, potentially to engage with them as new trading partners for the post-Brexit era. The most notable of these was a tour in 2019 to Morocco by the Duke and Duchess of Sussex and a tour of Cuba by Prince Charles and the Duchess of Cornwall, at the end of a tour of the Caribbean.

Conclusion

As mentioned above, the branding of the British royal family is dependent not only on staying relevant in a changing world, but also on staying aloof from overtly political acts. It is a delicate balancing act, and can also be undone by a sense that the royal family are being insensitive to the travails of their people. The cost of a baby shower in New York in honour of the first child of the Duke and Duchess of Sussex attracted highly critical headlines. As Elser (2019) observed: 'The image all of that crass, excessive consumption that went on in New York is a terrible one for an institution and family fighting to prove their relevance in 21st century Brexit Britain'. It occurred at a time of great uncertainty and ongoing political debate about Brexit, with the prime minister facing calls to resign and massive difficulties in negotiating a Brexit deal through Parliament. It was a jarring note, given the consistently positive public relations messages that had been delivered previously through royal events.

Whatever the outcome of the Brexit negotiations, it is clear that there is a role for royal events in supporting government endeavours to return to stability. Throughout Brexit, royal events have been important in two ways. First, they have been a mechanism for maintaining and reinforcing existing alliances, particularly with the Commonwealth countries and the United States. Second, they are increasingly used to develop new relationships and markets for Great Britain, in the post-Brexit world. Future research will need to continue to explore the role of royal events, given the volatility of the political scene as the situation unfolds.

References

Alomes, S. (1988) *Nation at Last? The Changing Character of Australian Nationalism 1880–1988*. Sydney: Angus & Robertson.
Bagehot, W. (1867) *The English Constitution* (2nd edn 1905). London: Kegan Paul.
Balmer, J.M.T., Greyser, S.A. and Urde, M. (2006) The crown as a corporate brand: Insights from monarchies. *Brand Management* 14 (1/2), 137–161.
BBC (2017) Duke and Duchess of Cambridge to visit Germany and Poland. BBC, 3 March. See https://www.bbc.com/news/uk-39156396 (accessed 5 March 2019).
Bramston, T. (2015) Queen's ex-aide William Heseltine reflects on her life and times. *The Weekend Australian*, 14 September. See http://www.theaustralian.com.au/news/inquirer/queens-exaide-william-heseltine-reflects-on-her-life-and-times/news-story/6df513f2330f0942e84418cf0855b5c4 (accessed 12 February, 2017).
Brunt, R. (1992) A 'divine gift to inspire'? Popular cultural representation, nationhood and the British monarchy. In D. Strinati and S. Wagg (eds) *Come on Down? Popular Media Culture in Britain* (pp. 285–301). London: Routledge.
Cannadine, D. (1983) The context, performance and meaning of ritual: The British monarchy and the 'invention of tradition', c. 1820–1977. In E. Hobsbawm and T. Ranger (eds) *The Invention of Tradition* (pp. 101–164). Cambridge: Cambridge University Press.
Cannon, J.A. (1987) *The Modern British Monarchy: A Study in Adaptation*. Reading: University of Reading.
Chace, J. (1973) The concert of Europe. *Foreign Affairs* 52 (1), 96–108.
Connors, J. (1993) The 1954 royal tour of Australia. *Australian Historical Studies* 25 (100), 371–382.
Dobransky, K. and Fine, G.A. (2006) The native in the garden: Floral politics and cultural entrepreneurs. *Sociological Forum* 21 (4), 559–585.
Elser, D. (2019) Meghan Markle's $348k baby shower breaks number one royal rule. *News.com*, 25 February. See https://www.news.com.au/entertainment/celebrity-life/royals/meghan-markles-348k-baby-shower-breaks-number-one-royal-rule/news-story/bdf8717b79c2aff244402e48f7ef5bfa (accessed 25 February 2019).
Feingold, R.P. (2009) Marketing the modern empire: Elizabeth II and the 1953–1954 world tour. *Antipodes* 23 (2), 147–154.
Hague, W. (2011) The Commonwealth is 'back at the heart of British Foreign Policy', 27 July. See https://www.gov.uk/government/speeches/the-commonwealth-is-back-at-the-heart-of-british-foreign-policy (accessed 28 January 2019).
Hardman, R. (2012) *Her Majesty: Queen Elizabeth II and Her Court*. New York: Pegasus.
Harrison, B. (1996) *The Transformation of British Politics 1860–1995*. Oxford: Oxford University Press.

Haseler, S. (2012) *The Grand Delusion: Britain After Sixty Years of Elizabeth*. London: I.B. Tauris.
Hobolt, S.B. (2016) The Brexit vote: A divided nation, a divided continent. *Journal of European Public Policy* 23 (9), 1259–1277.
Hobsbawm, E. (1983) Introduction: Inventing traditions. In E. Hobsbawm and T. Ranger (eds) *The Invention of Tradition* (pp. 1–14). Cambridge: Cambridge University Press.
Kuhn, W.M. (1996) *Democratic Royalism: The Transformation of the British Monarchy, 1861–1914*. Basingstoke: Macmillan.
Laing, J. and Frost, W. (2018) *Royal Events: Rituals, Innovations, Meanings*. London: Routledge.
Long, P. (2008) Introduction. In P. Long and N.J. Palmer (eds) *Royal Tourism: Excursions Around Monarchy* (pp. 1–25). Clevedon: Channel View Publications.
Lowe, J. (2016) Commonwealth Day: Why does it matter to the U.K.'s Brexit camp? *Newsweek*, 14 March. See https://www.newsweek.com/commonwealth-day-brexit-ukip-eu-436481 (accessed 28 January 2019).
McIntyre, W.D. (1991) *The Significance of the Commonwealth 1965–90*. Houndmills: Macmillan.
Nairn, T. (1988) *The Enchanted Glass: Britain and its Monarchy*. London: Radius.
Namusoke, E. (2016) A divided family: Race, the Commonwealth and Brexit. *Round Table* 105 (5), 463–476.
New Zealand Ministry for Culture and Heritage (2014) Royal tours. *Queen Elizabeth's Diamond Jubilee*, 11 July 2014. See https://nzhistory.govt.nz/politics/queen-elizabeth-jubilee/royal-tours (accessed 22 March, 2019).
Prince, Son and Heir: Charles at 70 (2018) [film] Directed by J. Bridcut. Arlington, VI: PBS.
Prochaska, F. (1995) *Royal Bounty: The Making of a Welfare Monarchy*. New Haven, CT: Yale University Press.
Richards, J. (2004) The coronation of Queen Elizabeth II and film. *The Court Historian* 9 (1), 69–79.
Royal Family (1969) [film] Directed by R. Cawston. London: BBC and ITV.
Royal Family (2019) 50 Facts about the Queen's coronation. See https://www.royal.uk/50-facts-about-queens-coronation-0 (accessed 5 January 2019).
Royal Household (2017) Christmas Broadcast 1953. *The Home of the Royal Family*. See https://www.royal.uk/christmas-broadcast-1953 (accessed 5 July 2017).
Specter, F. (2017) THIS is how much the royals EARNED for British tourism this year. *The Daily Express*, 20 November. See https://www.express.co.uk/travel/articles/882003/royals-earned-millions-british-tourism-industry-2017 (accessed 20 March 2019).
Spearritt, P. (1988) Royal progress: The Queen and her Australian subjects. In S.L. Goldberg and F.B. Smith (eds) *Australian Cultural History* (pp. 138–157). Cambridge: Cambridge University Press.
Strasdin, K. (2012) Empire dressing: The design and realization of Queen Alexandra's coronation gown. *Journal of Design History* 25 (2), 155–170.

6 Brexit and Tourism in Central and Eastern Europe: The Case of Poland

Sabina Owsianowska and
Magdalena Banaszkiewicz

Introduction

The discussion taking place around Brexit also concerns the issue of travel to and from the UK, which in this chapter is discussed with the example of Central and Eastern Europe. At the outset, it should be noted that the Central and Eastern Europe boundaries may be variously designated (see Banaszkiewicz *et al.*, 2017; Owsianowska, 2018); however, for the purpose of this chapter, we accept the division proposed by the World Tourism Organization (UNWTO). We mainly focus on mobility between Poland and the UK; nonetheless, the topic is presented within the context of other countries hidden behind the Iron Curtain up until the 1990s (Hall, 2017; Owsianowska & Banaszkiewicz, 2018). In addition to analysing statistical data which illustrate tourist flows, popular destinations and forms of tourism, the chapter aims to provide insight into the ways in which travel between the UK and Poland is promoted and described, on the basis of selected campaigns by the Polish Tourist Organisation (PTO) and printed and online travel guides. Significantly, we pay attention to the strategies of narrating 'Britishness' in the examined tourist media.

It is difficult to unambiguously write about Brexit and its consequences in a situation where nobody – including the British people and the politicians at the forefront of negotiations – has any idea how the process will proceed. Nevertheless, with this lack of surety in mind, we will attempt to discuss the potential consequences of Brexit in the context of an analysis of the Polish travel market in recent decades.

From the point of view of tourism in Central and Eastern Europe, the accession of Poland and other countries to the European Union (EU) in 2004 was a breakthrough. Political, economic and sociocultural transformation; the development of low-cost airlines and other travel facilitation; and the gradual changes in stereotypical images of post-socialist countries, also possible due to the development of tourism and promotional initiatives, is just one side of the coin. On the other side, there are the realities of economic migration to the UK and the growing diaspora, especially of Poles, which motivates visiting friends and relatives (VFR) tourism, back-to-roots travel and ethnic tourism (Kuzmicki & Wasilewska, 2015). In addition, we will discuss some of the consequences of the expansion of the tourism industry in Central and Eastern Europe destinations in relation to sustainability.

Tourism in Poland

The year 2017 brought an 8% growth in European tourism with 538 million international tourist arrivals, accounting for 40% of the world's total arrivals (The European Union Tourism Trends, 2018). The dynamics of tourist traffic in Central and Eastern Europe is not as spectacular as in Asia in terms of growth, and it is far from the position of leaders such as France or Spain in terms of tourist numbers in general, but we can see a systematic increase in the interest of tourists in this area. The European Union Tourism Trends (2018) report, by the European Commission and the UNWTO, makes a distinction within the Central and Eastern European subregion between so-called EU destinations and extra-EU destinations, but does not refer to Central Europe *per se*. According to the report, in 2016, Central and Eastern Europe recorded a 4% increase in arrivals (127 million) and a 6% increase in receipts (€48 billion), though arrivals to five countries (Bulgaria, Poland, Hungary, Romania and Croatia) grew even faster (at a rate of 8%, while receipts grew by 10%, reaching €29 billion) (The European Union Tourism Trends, 2018). In 2016, international arrivals to Poland increased by 4% to 17 million, while tourism earnings grew by 10% to €10 billion. This is connected to two significant events in 2016 – Wrocław, in western Poland, was one of the 2016 European Capitals of Culture, while Krakow, in southern Poland, hosted World Youth Day, an event organised for young people by the Catholic Church.

According to the World Economic Forum (2017), Poland was ranked 46th in the travel and tourism competitiveness index. Among the other countries of the region, Estonia was 37th, the Czech Republic 39th and Slovenia and Bulgaria were ranked slightly higher at 41st and 45th, respectively. However, it is important to note that Poland's position is systematically strengthening, improving from 58th position in 2009, to 49th in 2011, to 46th place in 2017. In ranking the extent of the effects generated by the tourist industry (expressed in real prices in terms of dollars) in the years 2005–2015, Poland was ranked 12th among the 28 EU member states, although it was only 25th in terms of the contribution of tourism to the country's gross domestic product (GDP).

Studies conducted by Statistics Poland, the chief government executive agency charged with collecting and publishing statistics related to Poland, show that the rate of development of tourism in Poland remains at a constant level. In 2017, 83.8 million foreigners (including 18.3 million tourists and 65.5 million day visitors) came to Poland (4.1%, 4.4% and 4.0% more than in 2016, respectively). Nearly half (48.2%) of the international arrivals were in the 35–54 age group, with people aged 25–34 also constituting a large group (18.8%). In 2017, expenditure by international tourists to Poland amounted to 56.7 billion PLN, which was 3.6% more than in 2016. In 2017, the largest number of tourists was recorded in the administrative areas or voivodships of Malopolska (5.1 million including 1.5 million foreigners) and Mazowieckie (4.9 million). It was 6.2 million in Warsaw and 5.3 million in Kraków. These two cities occupy the leading position for tourist numbers if the number of overnight stays is taken into account (Statistics Poland, 2017).

Tourism from the UK

Numbers collected by the border authorities for the Schengen area indicate that the highest numbers of visitors coming to Poland in 2017 were citizens of Ukraine, Belarus and Russia. Arrivals from the UK ranked fourth at 809,000 or 4.4% of all arrivals, an increase of almost 20% compared to 2016.

According to Statistics Poland (2017: 43–45), the British are the third group of international tourists (after the Germans [33.6%] and the Ukrainians [7.7%]) using accommodation facilities in Poland (7.6%), with the exception of Kraków, where British tourists are the largest number in terms of bed occupancy at 18.6% of the total. Their participation in the market increased significantly in 2004–2008, which is undoubtedly connected with Poland's accession to the EU.

Table 6.1 shows the main tourist destinations in Poland in terms of the share of accommodation provided to foreign tourists in total number of overnight stays in Krakow. In this city, British visitors are the largest

Table 6.1 Tourism in Poland 2017

City	Share of accommodation provided to foreign tourists in total number of overnight stays (%)	Germany (%)	Great Britain (%)	Israel (%)
Kraków	55.8	9.9	18.6	nd
Poznań	27	26.5	9.2	nd
Warsaw	41.5	nd	9.9	10.1
Wrocław	37.1	37.6	9.2	nd
Kołobrzeg	27.7	96	nd	nd

Source: Statistics Poland. See https://stat.gov.pl/obszary-tematyczne/kultura-turystyka-sport/turystyka/turystyka-w-2017-roku,1,15.html.

group of those who decide to spend at least one night, double the number of Germans who are usually the leaders among visitors using accommodation facilities in Poland.

However, taking tourists from countries that do not share a border with Poland into account, the largest percentage are visitors from the UK (4.7% of all tourists, excluding Polish residents who emigrated from the country), followed by tourists from Italy (2.9%), France (2.8%) and the United States (2.7%) (Statistics Poland, 2017: 103). Those arriving from the UK are also in the group with the highest monetary spend on a trip to Poland (following those from Germany and the United States). In 2017, tourists from the UK spent PLN 1.6 billion in Poland, which was 4.9% of the total spend of foreign tourists (Statistics Poland, 2017: 105).

The extensive image-related study (Lenczewska, 2010) carried out by TNS OBOP – one of the biggest Polish agencies specialising in market research and public opinion polling – before the European Football Championship in 2012, which was jointly organised by Poland and the Ukraine, showed that the British had little or no knowledge about Poland (72% of respondents gave this answer). Knowledge derived mainly from the media (TV, press, internet news) and Polish tourists coming to the UK. However, the general image of Poland was primarily positive and based on the positive valorisation of Poles themselves, thanks to the intensification of relationships between the British and Poles after the opening of the British labour market. According to popular ethnic stereotypes and prejudices (Kissel, 2000: 23–25) popularised on the internet in the tourism context (see e.g. Stay Poland), Poles do not know foreign languages, are lazy, do not smile, complain a lot, are intolerant, overly religious and abuse alcohol. However, as the TNS OBOP study proves, unlike the Austrians and the Dutch, who more often attributed features to Poland such as corruption, bureaucracy, crime and alcohol, the inhabitants of the UK perceived Poles as open, creative and courageous. The most frequently cited associations by the respondents in the cited research regarding Poles were immigrants of Polish descent who come to the UK for financial gain (23%), Second World War (15%) and Polish products (10%). Although, according to the respondents, Poland is rich in attractions and monuments (9%), it does not have an attractive climate (10%). However, more significant is that 66% of respondents could not name any places in Poland, demonstrating the very weak tourist brand that is Poland (Lenczewska, 2010). Poland is generally perceived as a country rich in terms of tourism potential, but in practice, it is mainly seen at best as good value for money or simply as a cheap destination.

Although no studies specify tourism motivations in relation to individual countries, it can be assumed that tourism from the UK is part of the general trend. The main goals of international tourists to Poland in 2012–2017 were: VFR (18%–40%), rest (20%–28%) and business trips

(19%–26%). In this case, tourism from the UK is strongly associated with the migration of Poles after the opening of the British labour market in 2004 as a result of Poland joining the EU. In 2003, only 69,000 living in the UK declared Polish citizenship. According to the British Office of National Statistics, at the end of 2017, a record number of Poles lived in the UK – more than 1 million, or 19 thousand more than the previous year, which means that Polish citizens constitute the largest national minority in the UK (ONS, 2018). In 2017, among the 18.3 million international tourist arrivals, 3.4 million were of Polish citizenship and 0.1 million were those who had a Pole's Card – a document confirming belonging to the Polish nation granted to people who do not have Polish citizenship (Polska Organizacja Turystyczna, 2017).

The increasing share of travellers from the UK to Poland stimulated the development of the transport industry. Data from the Civil Aviation Authority in Poland indicate that in 2018, approximately 7.9 million passengers were carried on routes between Poland and the UK, which is almost 22% of all Polish air traffic (Urząd Lotnictwa Cywilnego, 2019). In 2015, there were 6.2 million passengers, and a year later 7.1 million. A good example of the trend is the low-budget British carrier easy-Jet which, in the period 2004–2018, served over 7 million passengers between London and Kraków (London is actually the most popular city to arrive and depart from which most passengers between the UK and Poland travel – in 2017, there were 4.4 million passengers). Currently, Krakow Airport maintains regular flights with four of London's airports: Heathrow (British Airways), Gatwick (easyJet), Stansted (Ryanair) and Luton (easyJet, Wizz Air). Inexpensive flights run twice a day (Ryanair and easyJet) contributing to the rapidly increasing number of passengers served by Krakow Airport. In 2018, this number was 6.7 million passengers, while the current capacity of the airport is not much greater, at only 8 million passengers (Mw, 2019).

Direct flights to various cities in Poland are an important factor in the development of inbound tourism. The most popular tourist destinations among the British are cities attractive because of their historical value and their cultural and entertainment offerings. For several years, Warsaw, Krakow, Tricity (Gdansk-Gdynia-Sopot), Zakopane and Malbork have been in the leading position for heritage tourism. It is not surprising, therefore, that the 'leitmotif' of the Polish stand at the World Travel Market in London in 2018 was primarily urban tourism, nature and Polish cuisine. An attempt to diversify British interests in Poland's offerings via the promotion of rural areas became possible due to cooperation between Poland's Ministry of Sport and Tourism and the Ministry of Agriculture and Rural Development, thanks to which more emphasis was placed on agrotourism and the 'slow tourism' movement, which is generally becoming a more visible trend in European tourism (Polskie stoisko narodowe, 2019).

British Tourists in the Polish Tourist Organisation's Campaigns

Since 2000, the PTO, under the supervision of the Department of Tourism, has been promoting Poland as a country attractive for tourists. The promotion of Poland on the world stage is the responsibility of the PTO's foreign offices (ZOPOT), which operate in 14 countries: Austria, Germany, Sweden, the Netherlands, Belgium, the UK, Italy, France, Spain, the Ukraine and Russia, and outside of Europe in the United States, Japan and China. The geographical distribution of ZOPOT correlates with the most important countries for Polish inbound tourism – this applies to 9 out of 10 European markets and 4 out of 5 non-European markets (cf. OECD, 2016).

In June 2018, a new director in the form of Iwona Białobrzycka, a specialist in sales and tourism marketing, began a 33-month term of office for ZOPOT in London. As emphasised by the chief executive officer (CEO) of the PTO, Robert Andrzejczyk, after Białobrzycka's appointment:

> In London, we focus on the British interest in a wider offer than the one that is now utilised by them. Most often, they visit Polish cities, dropping by for the weekend. We want them to come for a longer duration – to take tours and undertake active tourism such as cycling or canoeing. Our representative in London must pay more attention to the Republic of Ireland. After the exit of The UK from the European Union, Poland may become more attractive to the Irish as a destination. (Frydrykiewicz, 2018)

This quote from the PTO director is a good introduction to the issues presented in this part of the chapter, focusing on tourist narratives.

Narratives play a key role in shaping tourists' experiences, which are the subject of reflection for researchers from various disciplines and fields of knowledge (Bruner, 2005; Dann, 1996; McCabe & Foster, 2008; Owsianowska, 2014; Tivers & Rakić, 2012). Although the main functions of narratives remain unchanged, the manner of constructing storylines is transformed due to the evolution of the language of new media and cyberculture, as well as the meaning of so-called 'narrative marketing' in the promotion of places and products. Notably, the analysis of tourism discourse not only concerns the credibility of information and persuasive effectiveness, but also refers to the content of the message infused with cultural values, signs and symbols. The images of British markers of national and ethnic identity and stereotypes are transferred in pre-, on- and post-trip media (Dann, 1996). Hereafter, we have taken into consideration the key initiatives of the PTO since 2011. They are part of a project of long-term activities which were undertaken during the Polish presidency of the EU in 2011 and the organisation – together with

Ukraine – of EURO 2012. These events provided an excellent opportunity to initiate a comprehensive promotional campaign and an attempt to create a new, recognisable image of the country (Image Research, 2011; Johann, 2014; Owsianowska, 2014, 2017).

In the campaign 'Poland. Move your imagination', British tourists are portrayed in a series of films entitled 'It happened in…', which centre on the key tourism destinations, mainly the cities of Krakow, Warsaw, Wrocław, Gdansk, Lodz and Poznan, and a region situated in north-eastern Poland – Mazury Lake District (Poland – Move your imagination, 2011). For example, one narrative is a romantic adventure featuring Sharon and Harrison, a middle-aged couple from the UK, visiting Krakow and celebrating the artistic and sophisticated dimensions of urban life. This creates an image of Krakow as a place for love and art. The goal of the campaign is to attract cultural tourists and it strictly corresponds with the official strategy of tourism development in Krakow. The city is presented as the best destination for visitors who prefer luxury, e.g. an exclusive hotel with a view of the Main Market Square in the centre of Krakow. The most remarkable monuments (e.g. Wawel Castle), museums and cultural events, such as a spectacle in the Juliusz Slowacki Theatre, are of interest to those who appreciate the arts and history. In the film 'It happened in Krakow', the whole city is like a huge stage with hospitable and friendly inhabitants who perform as mediators of tourists' experiences (Owsianowska, 2011).

British tourists are also presented in two subsequent PTO promotional campaigns entitled 'Poland. Feel invited' and 'Poland. Come and find your story' (Owsianowska, 2014, 2017). Bearing in mind the results of the research (Image Research, 2011) identified in the preceding discussion which indicated that Poland is not a well-known destination, the marketing campaigners chose an attractive and unconventional way to convince the inhabitants of Western European countries, mainly from the UK, France and Germany, to travel to Poland. Based on their research, PTO decided that promotion would be tailored to tourists in the 20–35 age range; for couples aged 35 and over, without children; for couples aged 55 and over; and for meetings, incentives, congresses and events (MICE) organisers. The main forms of tourism promoted were urban and cultural tourism events, nature-based visits, with Białowieza Forest in north-eastern Poland one of the must-see attractions, active tourism and business trips. Travellers share their adventures and encounters with people and places in a series of films and billboards. For the UK, these are Joe, Sheila, Kate, Rachael and James whose points of view, memories and emotions frame the stories. Fairy tales, romance literature and cinematic scenarios are designed to encourage tourists to visit the country. The images of Poland and Poles are shaped through travellers' embodied and polisensual experiences, for example in the creation entitled 'I've lost my head in Cracow. Polska by Joe' (Owsianowska, 2017).

Outbound Tourism to the UK

According to data from Poland's Ministry of Sport and Tourism, in 2017 Poles participated in 12 million foreign trips with at least one overnight stay. Foreign trips are predominantly (about 80%) long-term trips (5 days or more), and their main purpose of travel is usually for leisure and recreation. Foreign trips are also popular for VFR and for business purposes. As Statistics Poland indicates (Statistics Poland, 2017: 95), Poles definitely prefer trips to European countries compared to going outside of Europe (almost 92% of the entire travel market), most of which (2.35 million) in 2017 were trips to Germany, about 1 million to Italy and almost as many (971,000) to the UK. Other destinations include ones more traditionally related to summer recreation in 'warm countries': Croatia (almost 684,000), Greece (671,000) and Spain (662,000).

British statistics specify the distinctiveness of arrivals of Poles to the UK, showing that of the 1.8 million arrivals (not only tourists), over 640,000 were overland, thus, they were strictly connected with profit-making objectives based on the transport of produce (VisitBritain, 2019). In addition, business arrivals represented slightly more than 45% of the total market share (about 830,000). As mentioned above, the most popular reason for Poles' travelling to the UK is VFR (675,000, amounting to 37% of all border crossings), which is the most noticeable consequence of emigration in recent years. Visits strictly for leisure purposes (sightseeing, recreation, etc.) occupy third place with a market share of 12% (222,000) (VisitBritain, 2019).

According to the VisitBritain (2019) portal: 'Views of Britain tend to be positive: Polish respondents ranked the UK second out of 50 countries in 2017. They rate Britain highly for its contemporary culture, sports and vibrant cities. Music and films are the cultural products or services most associated with Britain, closely followed by sports and museums'.

From the point of view of the British tourist market, although the average stay of a Pole in the UK is relatively short (1–3 nights – 45.7% of the market share, 4–7 nights – 26%), what is important, mainly due to the dominance of the VFR sector, is that it occurs regardless of the time of year, and due to the price attractiveness of the low season, the strongest quarters for visits are the first and the last quarter of the year. Also, due to the specificity of inbound motivation, 60% of those arriving in the UK use private accommodation, 13% choose short-term rental of flats/houses or other forms of accommodation and only 10% opt for hotels or guest houses.

An interesting niche when it comes to Polish tourists in the UK is language courses based on the idea of linguistic and cultural immersion. This offering is particularly popular among young people – mainly teenagers and first-year university students, for whom a holiday combined with intensive English language learning is an attractive opportunity to develop competences relevant to further education and career planning. Often, participation in such courses is meant as the first step towards later study in the UK (before Poland joined the EU, the number of Poles studying at Oxford University remained at about 20 students a year, while in 2017, it had reached 250, while the number of all students of Polish descent in the UK has stabilised at around 6,000 individuals [Bagniewska, 2018]). Not surprisingly, these educational travels enable the acquisition of more knowledge about the UK and its residents. As for the other types of tourism, traditional and modern guidebooks (including participant-generated forms, e.g. travel blogs) are among the basic sources of information. Consequently, as they affect tourism imaginaries, the next part of the chapter discusses narratives of travel guidebooks.

Understanding the United Kingdom: Travel Guidebook Narratives

According to Graham Dann (1996), who highlights the tautological character of the language of tourism, well-known images from popular guidebooks circulate in other media, including travellers' testimonies, e.g. in participant-generated content available online (Dann, 2011). We have analysed two publications translated from English (e.g. *Great Britain – Key Guides* [Bamber *et al.*, 2007] and *London – Marco Polo* [Weber, 2017]) and publications by Polish authors (e.g. *Anglia. Przewodnik ilustrowany* [*England. An Illustrated Guidebook*], *Anglia i Walia. Praktyczny przewodnik* [*England and Wales. A Practical Guidebook*] and *Szkocja. Przewodnik* [*Scotland. A Guidebook*]). One of the most popular guidebooks is the 'Eyewitness Travel Guides' 'Great Britain'. A Polish language version was published in 1996. It has since been updated several times with the latest edition appearing in 2014 (Barnard *et al.*, 2014). Its portrayal of the country and its inhabitants may be considered as a point of reference for imaginaries associated with the UK.

From our analyses, we identified the most common topics in tourism media about the UK. The main topics are the royal family; ethnic diversity and regional distinctiveness; multiculturalism and the openness of the society; five-o'clock tradition for having tea; culinary heritage; sports and other hobbies; fashion; language(s); and religion(s). The British[1] are portrayed as great sports enthusiasts and among the popular sports are football, rugby, cricket and golf. Their favourite forms of activity include fishing, walking, hiking in national parks and gardening (Barnard *et al.*, 2014: 626–628), as well as regular visits to the pub. Other leisure-time activities include live music, various traditional games (skittles, dominoes, darts) and bicycle races (Barnard *et al.*, 2014: 37, 608–611). In summer, activities mainly take place outdoors, thus cafes, concerts, village fairs, crowded beaches and picnics in city parks are described. The authors present festivals, carnivals, fairs, tournaments, exhibitions and other events, all organised primarily for tourists or those related to official ceremonies – from the Chelsea Flower Show to St Patrick's Day and the Queen's birthday celebrations.

Tourists who want to know the customs of the inhabitants are advised to visit rural areas. The authors of the guidebook invoke an opposition between modernity and tradition (cf. Prieto Arranz, 2006a, 2006b) and write: 'Despite the development of cities over the last two centuries, the English countryside is still flourishing. Almost three quarters of the UK area is agricultural land' (Barnard *et al.*, 2014: 17–21) and 'The unique atmosphere results from the combination of the culture of industrial cities with rural tradition. Spas, historic villages and magnificent residences fit in perfectly with the landscape transformed as a result of industrial development' (Barnard *et al.*, 2014: 317).

The landscape of the country is not only beautiful, picturesque and inspiring, but it is also diverse. The reader learns that the nature of residents of different regions of the UK (Andrews, 2017) was shaped by the island's location, for example 'the English do what they have done for ages: they adapt their own traditions to the influences of other cultures, leaving essential elements of their own life and national character intact' (Barnard *et al.*, 2014: 19; cf. Lipiński, 2006).

The descriptions in the guidebook show cultural distinctiveness and numerous festivals that refer to the traditions of ethnic groups, cultivating their music, artisanship and languages:

> For such a small country, the United Kingdom consists of astonishingly diverse regions, the inhabitants of which have maintained their identity to this day. […]. The Welsh and Scottish languages […] have survived up until our times largely thanks to regional radio and television stations. In North England and the West Country, English has rich dialects and accents, and these areas also cultivate their own craft, art, architecture and cuisine. (Barnard *et al.*, 2014: 17–21)

Cuisine is the sphere of life in which attachment to traditional tastes is visible on the one hand, while the consequences of internationalisation and globalisation can be found on the other. Gastronomic heritage is the subject of assessment: 'The object of mockery, lack of imagination, a limited number of ingredients, unsophisticated preparation – typically English food means "simple meals" prepared at home or regional dishes' (Barnard *et al.*, 2014: 473). Some authors, however, signal changes in British cuisine: 'this favourable metamorphosis can be attributed to the influx of chefs from other countries and the influence of culinary traditions of other nations', e.g. inspirations from Thai, Mexican, Turkish, Tuscan or Indian cuisine or Hungarian, Polish and Caribbean restaurants (Barnard *et al.*, 2014: 473–474).

The above-mentioned signs of Britishness connected with everyday life, history and architecture, cultural heritage ('Culture first!'), contemporary art and museums, centre of commerce and banking, vibrant lifestyles, outstanding nature and rural traditions are the centre of interest for outbound tourism from Poland to the UK. On the one hand, places such as London, Stonehenge, Greenwich and Liverpool, to name a few, are listed as must-see in the homeland of Shakespeare, Sherlock Holmes, James Bond and Harry Potter. In fact, it is a country 'so different from Europe', we read in the guidebook (Dylewski & Czub, 2015a; cf. Lipiński, 2006). On the other hand, a presentation of British tourist attractions in an internet travel article, closes as follow:

> Ascot, Oxbridge, tea and a cucumber sandwich. Good Old England... Remember, however, that this Great Britain from our dreams and novels is a hermetic and extremely bourgeois creation... You will also get to know the real 'Englishmen' on the Main Market Square in Krakow, when they are squatting tenement houses... Or on merciless Martin Parr's photographs. Undeterred by this, I am reaching for a strawberry with whipped cream and champagne, to feel like at Wimbledon. (Kozak, 2008)

Recently published travel guidebooks include commentaries on political issues, e.g. Brexit (Dylewski & Czub, 2015a: 18–20; Kasperczyk, 2019: 7). The authors try to highlight and explain the social tensions and divisions which have been expressed in opinions towards the EU in the referendum. They refer to the imperial and colonial past of the UK and to postcolonial processes which resulted in the development of a multicultural society (Bamber et al., 2007; Dylewski, 2015a, 2015b; Kasperczyk, 2019; Weber, 2017; see also Davis, 1999; Rosiak, 2018). However, the increasing importance of identity politics has placed nationalist sentiments in the spotlight in the ongoing discussions about Brexit in the UK (e.g. Jenkins, 2016; Okri, 2018; Pappas, 2019; Świeszewski, 2019). In the opinion of the authors of the guidebook, it might be a response to frustration and nostalgia, because arguably: 'Since the creation of Great Britain England has paradoxically lost the most. [...] In the crucible of the United Kingdom, the English, although they constitute 80% of British society, get lost, being together with Scots, Welsh people and numerous descendants of former immigrants simply British' (Dylewski & Czub, 2015a: 18).

Nowadays, English self-awareness and identity are often demonstrated by, for example, flags decorating houses and cars during sports events or celebrations of St George's Day. The role of institutions protecting national cultural heritage (e.g. English Heritage) or a call for a separate English Parliament are also noticed in the chapter entitled 'Englishmen', where the question concerning identity also appears: 'Englishman or Britishman?' (Dylewski & Czub, 2015a: 20; on national identity in the context of tourism, see e.g. Andrews, 2011, 2016, 2017; Edensor, 2004; Palmer, 2005). The genesis of conflicts within British society, briefly described in the guidebooks, has the goal of trying to make it easier to understand the rise in popularity, in some quarters, of voices supporting the future of the UK outside the EU.

Conclusion

To conclude this chapter, discussing Brexit in the context of inbound and outbound tourism in Poland and Central and Eastern Europe, we briefly refer to the issues of sustainability. Dynamic and uncontrolled development of the tourist industry in the transitory period after the fall of the Iron Curtain is associated with numerous dysfunctions. For cities in the region, experiencing a tourism boom thanks to low-cost airlines, the competitive prices of services and their popularity (e.g. visiting new EU member states or the organisers of EURO 2012), resulted in what is called overtourism, gentrification and other costs (Borkowski *et al.*, 2017; Kruczek, 2018; Tracz *et al.*, 2019). The popularity of short trips by British men, searching for cheap beer, fun and sex, is an example of controversial forms of tourism in Central Europe. The region is one of the top destinations for stag tourism, after the UK and Ireland, the Iberian Peninsula and the Baltic states (Iwanicki *et al.*, 2016; Thurnell-Read, 2011, 2012). The most popular countries are the Czech Republic, Germany, Poland, Slovakia and Hungary, and the highest-ranked cities are Bratislava, Budapest, Krakow, Prague, Riga, Tallinn, Berlin, Brno, Sofia and Vilnius. The next most popular destinations include Gdansk, Warsaw and Wrocław in Poland; Ljubljana in Slovenia; and Zagreb in Croatia (Iwanicki *et al.*, 2016: 21–22).

Obviously, the most attractive historical cities derive economic profits from weekend arrivals for entertainment, but both service providers and residents report numerous problems. Therefore, in the centre of Krakow one can find both the 'Little Britain' pub in front of the Mariacki church and notices on the doors of several restaurants that 'stag parties are not welcome'. Tracz *et al.* (2019) studied the perception of tourism development by the inhabitants of Krakow. They reported that the most undesirable consequences of tourism included congestion and traffic restrictions in the city centre, an increase in the number of pubs and nightclubs, drunkenness, brawls and participation in sex tourism. Service providers, who also complained about damage to hotels and aggressive behaviour, drew attention to the fact that party-style tourism displaces cultural tourism. They argue that it has a negative impact on the image of Krakow as a United Nations Educational, Scientific and Cultural Organisation (UNESCO) World Heritage city and on a sense of place (Tracz *et al.*, 2019). This is a problem affecting many historical cities in this part of Europe and a challenge for local authorities and stakeholders in the changing and globalised world.

As far as international tourism in a post-Brexit era is concerned, different scenarios have been created (Jacobs, 2018; Pappas, 2019; Vickers,

2019). The debate that flared before the 2016 referendum and which is still not decreasing in importance reflects the cultural and political processes that affect such issues as national and common European heritage, hopes and fears related to integration within the EU and the impact of populist movements on societies. Although these topics go beyond the phenomenon of travelling, contemporary tourism and its narratives take part in the promotion of images of people and places, reinterpreting history, building and strengthening group identity and unity. The examples discussed in this chapter seem especially significant within the context of travels to and from the UK because of the waves of migration from Central and Eastern Europe to the country.

Note

(1) We are aware that English and British are often used interchangeably. However, we quote guidebooks or travel articles to show, inter alia, how the discourse about the UK and its citizens is created in the language of tourism.

References

Andrews, H. (2011) *The British on Holiday*. Bristol: Channel View.
Andrews, H. (2016) Brits abroad: Stereotype or media hype? *The Conversation*, October 7. See http://theconversation.com/brits-abroad-stereotype-or-media-hype-63973 (accessed 1 August 2019).
Andrews, H. (2017) Touring the regions. (Dis)Uniting the kingdom on holiday. *Journeys* 18 (1), 79–106.
Bagniewska, J. (2018) *Polscy naukowcy w Wielkiej Brytanii*: 'Keep calm and curie on' [Polish scientists in the UK: 'Keep calm and curie on'], blog. See https://joannabagniewska.com/2018/03/03/polscy-naukowcy-w-wielkiej-brytanii-keep-calm-and-curie-on/ (accessed 3 March 2018).
Bamber, J., Bennett, O., Follet, C., Gerrard, M., Locke, T., Staddon, J., Weston, H., White, J., Wiliams, N. and Winpenny, D. (2007) *Wielka Brytania [Key Guide Britain]*. Warsaw: Hachette Livre Polska.
Banaszkiewicz, M., Graburn, N. and Owsianowska, S. (2017) Tourism in (post)socialist Eastern Europe. *Journal of Tourism and Cultural Change* 15, 109–121.
Barnard, J., Catling, Ch., Clough, J., Hunt, L., Phillimore, P., Symington, M. and Roger, T. (2014) *Wielka Brytania [Great Britain]*. Warsaw: Wydawnictwo Wiedzy i Życia.
Borkowski, K., Grabiński, T., Seweryn, R., Wilkońska, A., Mazanek, L. and Grabińska, E. (2017) *Ruch turystyczny w Krakowie in 2017 [Tourist Movement in Krakow in 2017]*. Krakow: Urząd Miasta Krakowa.
Bruner, E. (2005) The Role of Narrative in Tourism. On Voyage. New Directions in Tourism Theory (Conference paper). University of Berkeley, 2–5 October 2005.
Dann, G. (1996) *The Language of Tourism: A Sociolinguistic Perspective*. Wallingford: CABI.
Dann, G. (2011) Take me to the the Hilton: A paradigm of the language of tourism. *Folia Turistica* 25 (1), 23–43.
Davis, N. (1999) *Wyspy. Historia [Islands. History]*. Kraków: Znak.
Dylewski, A. and Czub, K. (2015a) *Anglia i Walia. Przewodnik praktyczny [England and Wales. A Practical Guidebook]*. Bielsko-Biała: Wydawnictwo Pascal.
Dylewski, A. and Czub, K. (2015b) *Anglia. Przewodnik ilustrowany [England. An Illustrated Guidebook]*. Bielsko-Biała: Wydawnictwo Pascal.

Edensor, T. (2004) *Tożsamość narodowa, kultura popularna i życie codzienne* [National Identity, Popular Culture and Everyday Life]. Kraków: Wydawnictwo UJ.
Frydrykiewicz, F. (2018) POT ma czterech stuprocentowych dyrektorów ZOPOT-ów [PTO has four hundred percent directors of ZOPOT]. *Rzeczpospolita*. See http://turystyka.rp.pl/artykul/1367770.html?print=tak&p=0 (accessed 6 June 2018).
Hall, D. (ed.) (2017) *Tourism and Geopolitics: Issues and Concepts from Central and Eastern Europe*. Wallingford: CABI.
Image Research (2011) *Badania wizerunkowe Polski i polskiej gospodarki w krajach glownych partnerow gospodarczych* [Image research of Poland and Polish economy in the countries of main economic partners]. Report commissioned by the Ministry of Economy. Warsaw: Ageron Polska.
Iwanicki, G., Dluzewska, A. and Smith Kay, M. (2016) Assessing the level of popularity of European stag tourism destinations. *Quaestiones Geographicae* 35 (3), 15–29.
Jacobs, F.B. (2018) *The EU after Brexit: Institutional and Policy Implications*. Basingstoke: Palgrave Macmillan.
Jenkins, S. (2016) Blame the identity apostles – they led us down this path to populism. *The Guardian* 1 December. See https://www.theguardian.com/commentisfree/2016/dec/01/blame-trump-brexit-identity-liberalism (accessed 1 August 2019).
Johann, M. (2014) The image of Poland as a tourist destination. *European Journal of Tourism, Hospitality and Recreation* 4, 143–161.
Kasperczyk, K. (2019) *Szkocja. Przewodnik* [Scotland. A Guidebook]. Pruszków: Oficyna Wydawnicza 'Rewasz'.
Kissel, N. (2000) *Passport Poland: Your Pocket Guide to Polish Business, Customs & Etiquette*. Navato, CA: World Trade Press.
Kozak, A. (2008) Co oznacza brytyjskość? [What does Britishness mean?]. See https://kobieta.wp.pl/co-oznacza-brytyjskosc-5982067085948033a (accessed 21 May 2019).
Kruczek, Z. (2018) Turyści vs mieszkańców. Wplyw nadmiernej frekwencji turystów na proces gentryfikacji miast historycznych na przykładzie Krakowa [Tourists vs. residents. The influence of excessive tourist attendance on the process of gentrification of historic cities on the example of Krakow]. *Turystyka Kulturowa/Cultural Tourism* 3, 29–41.
Kuzmicki, M. and Wasilewska, M. (2015) The image of Poland as a tourist destination in the country's main economic partners. *Annales Universitatis Mariae Curie-Skłodowska* 43 (3), 91–98. doi: 10.17951/h.2015.59.3.91.
Lenczewska, I. (2010) *Wizerunek Polski i Polaków. Główne wnioski z badań* [The image of Poles and Poland. Main research findings]. TNS OBOP. See https://msit.gov.pl/download.php?s=1&id=12265 (accessed 6 May 2019).
Lipiński, W. (2006) *Dzieje kultury brytyjskiej* [The History of British Culture]. Warsaw: Wydawnictwo Naukowe PWN.
McCabe, S. and Foster, C. (2008) The role and function of narrative in tourist interaction. *Journal of Tourism and Cultural Change* 4 (3), 194–215.
OECD (2016) OECD Tourism Trends and Policies 2016. Poland. doi: https://dx.doi.org/10.1787/tour-2016-32-en.
Okri, B. (2018) How to combat the populism that gave us Brexit? Active citizenship. *The Guardian* 30 January. See https://www.theguardian.com/commentisfree/2018/jan/30/populism-brexit-citizens-political-responsibility (accessed 1 August 2019).
ONS (2018) Population of the UK by country of birth and nationality: 2017. See https://www.ons.gov.uk (accessed 9 May 2019).
Owsianowska, S. (2011) Tourism promotion, discourse and identity. *Folia Turistica* 25 (2), 231–248.
Owsianowska, S. (2014) Stereotypes in tourist narrative. *Turystyka Kulturowa* 4, 1–24.
Owsianowska, S. (2017) Come and find your (love) story: Remaking the image of Poland as a tourist destination. *Via. Tourism Review* 11–12. See https://journals.openedition.org/viatourism/1775 (accessed 28 May 2019).

Owsianowska, S. (2018) Mediating Central and Eastern Europe in tourism discourse. In S. Owsianowska and M. Banaszkiewicz (eds) *Anthropology of Tourism in Central and Eastern Europe: Bridging Worlds* (pp. 143–161). Lanham: Rowman & Littlefield.

Owsianowska, S. and Banaszkiewicz, M. (eds) (2018) *Anthropology of Tourism in Central and Eastern Europe: Bridging Worlds*. Lanham: Rowman & Littlefield.

Palmer, C. (2005) An ethnography of Englishness: Experiencing identity through tourism. *Annals of Tourism Research* 32 (1), 7–27.

Pappas, N. (2019) UK outbound travel and Brexit complexity. *Tourism Management* 72, 12–22.

Poland – Move your imagination (2011) See https://www.youtube.com/watch?v=fFc7LmQqSU0 (accessed 21 May 2019).

Polska Organizacja Turystyczna (2017) *Turystyka przyjazdowa w latach 2012–2017* [Inbound tourism in 2012–2017]. See https://zarabiajnaturystyce.pl/obserwatorium-turystyki/ruch-turystyczny/turystyka-przyjazdowa/ (accessed 6 May 2019).

Polskie stoisko narodowe (2019) See https://www.pot.gov.pl/pl/nowosci/polecane/premiera-polskiego-stoiska-narodowego-na-wtm-londyn-2018 (accessed 6 May 2019).

Prieto Arranz, J.-I. (2006a) BTA's Cool Britannia: British national identity in the new millennium. PASOS *Revista de Turismo y Patrimonio Cultural* 4 (2), 183–200.

Prieto Arranz, J.-I. (2006b) Rural, white and straight: The ETC's vision of England. *Journal of Tourism and Cultural Change* 4 (1), 19–52.

Mw (2019) Rekordowy rok na Kraków Airport [Record year at Krakow Airport]. *Gazeta Wyborcza*, 3 January. See http://krakow.wyborcza.pl/krakow/7,44425,24328370,rekordowy-rok-na-krakow-airport-ponad-6-7-mln-obsluzonych-pasazerow.html (accessed 3 January 2019).

Rosiak, D. (2018) *Oblicza Wielkiej Brytanii. Skąd wziął się Brexit i inne historie o Wyspiarzach* [*The Faces of Great Britain. The Sources of Brexit and Other Stories about the Islanders*]. Wolowiec: Wydawnictwo Czarne.

Statistics Poland (2017) Tourism in Poland 2017. See https://stat.gov.pl/obszary-tematyczne/kultura-turystyka-sport/turystyka/turystyka-w-2017-roku,1,15.html (accessed 9 May 2019).

Stay Poland. See https://www.staypoland.com/poland/stereotypes-prejudices-of-poland/ (accessed 28 May 2019).

The European Union Tourism Trends (2018) See https://ec.europa.eu/growth/tools-databases/vto/content/2018-eu-tourism-trends-report (accessed 26 April 2018).

Świeszewski, A. (2019) Brytyjscy artyści komentują brexit poprzez sztukę [British artists are commenting Brexit through art]. *Polityka* 23 June. See https://www.polityka.pl/tygodnikpolityka/kultura/1790670,1,brytyjscy-artysci-komentuja-brexit-poprzez-sztuke.read (accessed 1 August 2019).

The World Economic Forum (2017) The Travel and Tourism Competitiveness Report 2017. See http://reports.weforum.org/travel-and-tourism-competitiveness-report-2017/ (accessed 9 May 2019).

Thurnell-Read, T. (2011) Off the leash and out of control: Masculinities and embodiment in Eastern European stag tourism. *Sociology* 45 (6), 977–991.

Thurnell-Read, T. (2012) Tourism place and space: British stag tourism in Poland. *Annals of Tourism Research* 39 (2), 801–819.

Tivers, J. and Rakic, T. (eds) (2012) *Narratives of Travel and Tourism*. London: Routledge.

Tracz, M., Bajgier-Kowalska, M. and Wójtowicz, M. (2019) Przemiany w sferze usług turystycznych Krakowa i ich wpływ na percepcję turystyki przez mieszkańców [Transformation in the tourist services and their impact on the perception of tourism by the residents of Krakow (Poland)]. *Studies of the Industrial Geography Commission of the Polish Geographical Society* 33 (1), 89–105. doi: 10.24917/20801653.331.13.

UNWTO See http://europe.unwto.org/ (accessed 28 March 2019).

Urząd Lotnictwa Cywilnego (2019) *Bardzo dobre wyniki przewozów pasażerskich w transporcie lotniczym w Polsce w 2018 roku* [Very good results of passenger air transport in Poland in 2018]. See http://www.ulc.gov.pl/pl/247-aktualnosci/4580-bardzo-dobre-wyniki-przewozow-pasazerskich w-transporcie-lotniczym-w-polsce-w-2018-roku (accessed 15 April 2019).

Vickers, B. (2019) Implications of Brexit. In T.M. Shaw, L.C. Mahrenbach, R. Modi and Y.-C. Chong (eds) *The Palgrave Handbook of Contemporary International Political Economy* (pp. 283–299). Palgrave Handbooks in IPE. See https://doi.org/10.1057/978-1-137-45443-0_18 (accessed 21 May 2019).

VisitBritain (2019) See https://www.visitbritain.org/sites/default/files/vb-corporate/markets/visitbritain_marketprofile_poland.pdf (accessed 21 May 2019).

Weber, B. (2017) Londyn. Przewodnik Marco Polo z mapą [*London. A Marco Polo's Guidebook with a map*]. Ostfildern: Mairdumont GmbH & Co.

7 From Duty Free to Benidorm: British Tourists in Spain in an Age of Brexit

Mark Casey

Introduction

During the 1950s, Western Europe witnessed the arrival of the modern package holiday – a lifestyle and leisure product that would significantly transform the cultures, geographies and economies of Europe and the lives of its people on a scale which would have been unimaginable to its originator, Vladimir Raitz. In 1952, Corsica was Raitz's initial site for this new undertaking, but it was his foray into Spain in 1954 that would make the country the most successful destination in terms of tourist numbers, with the country eventually emerging as the number one overseas holiday destination for British tourists, along with becoming home to the largest British expatriate community. During the 1950s and 1960s, UK travel companies worked alongside the Spanish authorities and local governments to identify coastal sites within Spain which could be transformed from often impoverished fishing communities into tourist resorts, providing a new economic model for local residents. To attract the growing middle class in Northern Europe, the Spanish authorities and travel companies developed a mass, low-cost tourism model which could be easily adopted across multiple sites throughout Spain (Valenzuela, 1988). Spanish coastal villages such as Torremolinos, Fuengirola, Benidorm, Magaluf, Marbella and Salou were identified as sites that could be developed quickly within a developmental model that would utilise and draw upon Spain's natural resources, for example the sunshine, its beaches and the Mediterranean sea, to attract Northern European tourists.

Prior to the arrival of mass tourism there were just over 1 million overseas visitors to Spain in 1951, exploring a country coming to terms with its brutal civil war and the consequential emergence of the Franco regime, and struggling with large-scale poverty among its general population (Turner & Ash, 1975). Within 30 years and by the early 1970s, Spain had significantly increased its number of overseas visitors to a staggering

34.6 million (Prieto-Arranz & Casey, 2004: 69). With the majority of these visitors arriving from Northern Europe and heading towards the increasingly large coastal resorts, Spain had now become the most important site in the world for the package holiday revolution (Bray, 2001). By the early 21st century, 13 million British tourists were visiting Spain annually, placing the UK as Spain's largest and most important international market, with the country overall witnessing a dramatic 4000% increase in the number of international arrivals since the mid-1950s (Bote & Sinclair, 1996: 65–66). It is possible to argue that these staggering tourist numbers, alongside the seven decades that British tourists have been consuming Spain's mass tourism model, have placed the country, its climate, cultures and people at the centre of the popular imagination of the British tourist and wider British media.

From *Duty Free* to *Benidorm*

Drawing from Barker (1999: 29), it is possible to argue that across the late 20th century and even into the 21st century, television remains 'the major communicative device for disseminating those representations which are constitutive of (and constituted by) cultural identity'. Prieto-Arranz and Casey (2014: 68) go on to explain that such an understanding of television's part in communicating identities and cultural representations has 'prompted studies on the role played by television drama in the (re)construction of national identities, thus pointing to the direct or indirect narration of the nation'. Although television is powerful in representing and constructing the identities of nations and their people, it is important to remember, especially in an age of large-scale university education, mass travel and the internet, that signs present in television programmes may have a dominant meaning but they will not always be passively accepted by viewers or understood in the way intended by programme producers (Schroder, 2007; Spencer, 2011). Images and stories presented by television programmes offer a limited insight into people, places and geographic settings, and as well as witnessing what is present and visible, it is also important to investigate who or what is absent. In doing so, it is possible to understand how television can make, recreate and deconstruct the identities of places and people in the popular imaginary and imagery.

The relationship between television and the viewer can be theorised as correlative, where dominant ideologies can be dispersed among the viewing public while also allowing an emergence of new ways of thinking and living within society, one in which the viewer consumes and reacts to the images and storytelling they have witnessed (Hall, 1980; Spencer, 2011). Many television programmes, even fictional programmes such as those which this chapter draws from, will offer viewers a 'lived experience', where familiar settings may be used, such as an airport or a resort, creating a relationship between the viewer and places and events that the viewer then perceives to be relevant to their own lives or experiences (Lewis, 2004). In light of this, the chapter will draw from and utilise two British television programmes, *Duty Free* (1984–1986) and *Benidorm* (2007–2018), in theorising the role of nationality and social class in creating a 'them' and 'us' binary between the British tourist and the local Spanish population. In doing this, the chapter will argue that in programmes such as *Duty Free* and *Benidorm*, tourism may be understood and presented as an act that divides visitors and locals, rather than uniting them in a shared European identity, one that maps onto wider notions of 'Britishness' and 'otherness' in the era of Brexit.

Duty Free was a television comedy that ran for three series in the mid-1980s on British television during a time when relations between Britain's then Conservative prime minister, Margaret Thatcher,[1] and the European Union (EU) were strained after a 'five year war of attrition over the EEC[2] budget', one in which Thatcher would not agree to the economic terms or the financial payments requested by the then EEC from the UK (Brown, 1984). For Fontana and Parsons (2015: 90), Thatcher's hard-line stance against the EEC would 'personally set Britain on a far more anti-European path than it was otherwise likely to take'. *Duty Free* was set in a fictional hotel in the upmarket Spanish resort of Marbella. The series focused upon two white heterosexual British couples, David and Amy Pearce and Robert and Linda Cochran. Although much of the programme was focused on the suggested sexual affair between David and Linda, there were two other consistent themes in the series that were central for maintaining humour. The first of these themes was the visible differences in politics and social class of the two main couples, working-class and Labour Party[3] voting David and Amy and middle-class and Conservative Party voting Robert and Linda. It is possible to argue that the distinct differences in the socioeconomic reality of each couple and their political allegiances epitomised the complex economic, political and regional realities of 1980s Britain. Robert and Linda can be understood as the winners of Thatcher's economic and neoliberal policies of the 1980s, they are economically wealthy, globally mobile, while being conservative in manner and in their political and social outlook on the world, a world in which Britain is presented as superior in its relationship with EU countries such as Spain.

However, David and Amy, although also staying in the same luxury hotel, are the economic losers of the 1980s. They are from the declining industrial lands of northern England; David has lost his factory-based job and is using his redundancy to fund the holiday. They embody their working-class identity through their northern English accents and, in particular, through Amy's inability to pass herself off as middle class through the way she moves, dresses and speaks that places them as 'out of time and place' in Thatcher's Britain. Jenson (2012) argues that the classed position of the working class can be identified through its juxtaposition with the middle and upper classes, particularly through their habitus. This can be seen in the way that the different classes stand, move, dress and carry themselves and in their ways of speaking[4] (see Lawler, 2014), with such traits repeatedly present in *Duty Free* which signals to the viewer the differences between and within each couple. The other key theme that runs throughout *Duty Free* is the distinct, inferior and suspect 'otherness' of Spain, its people and culture, which is embodied in the hotel's waiter 'Carlos' who is often presented to the other characters and viewer(s) at home as confused, unsure and as somewhat 'simple' in comparison to the internationally mobile British tourist. This will be discussed later in the chapter.

Benidorm was a British television comedy that ran from 2007 to 2018, airing at a time of significant global recession and increasingly strained relations between Britain and the EU. This was a time when anti-EU Conservative politicians forced the then UK Conservative prime minister, David Cameron, to commit to a referendum on exiting the EU if he won another term of office, alongside the powerful rise of Nigel Farage's UK Independence Party (UKIP) (see Fontana & Parsons, 2015; Parker & Flemming, 2014). Unlike *Duty Free*, *Benidorm* is not set in an upmarket resort, but in the working-class orientated Spanish resort of Benidorm, with Spain and the wider resort initially presented to the viewer as familiar and known. Various characters in the programme repeatedly compare Benidorm to the most famous of British working-class resorts 'Blackpool', but 'with added sun'.

The central characters of the show are the Garvey family, who are presented as economically and culturally sitting somewhere between the white working class and the growing 'chav class'[5] to be found in northern Britain (Jones, 2011; Nayak, 2006). The Garvey family consist of Mick, the father, who although likeable, is unable to hold down a job; Janice his wife who tries hard to manage the competing demands of her family; and their children Michael, who is initially a hyperactive young boy and (Chan)telle, an overweight teenage girl and single mother to mixed-race baby 'Coolio'. The Garvey's extended family includes Janice's mother Madge and her new husband Mel, a small-scale successful working-class entrepreneur. Although presented as loving, the family are chaotic, unruly, lazy and, at times, racist, sexist and homophobic in their encounters with other British tourists and Spanish locals. In a similar vein to David and Amy in *Duty Free*, the Garveys embody and perform their working-class identities through their cheap clothes, sunburnt bodies and dislike for 'foreign' food; they are the economic losers in a post-industrial recession Britain. In particular, the northern English accents of the Garveys present them to the viewer as working class in their unsophisticated accents and ways of speaking and for Allen and Mendick (2012: 467), 'particular [northern] English regional identities and accents are marked as working class', often against other regional accents, such as those from the southern home counties in the UK, which are often positioned as middle or upper class (also see MacLure, 2003).

The Spanish Other

The imagery, language and encounters presented in both *Duty Free* and *Benidorm* position Spain and the Spanish workers and/or residents as key characters in each programme, often in roles akin to the pantomime villain (see Lipton, 2008). Although initially it appears that *Duty Free* positions Spain and its people as 'exotic' and 'foreign', and *Benidorm* presents the spaces and places of the resort as familiar, it is clear that in both programmes Spain and its people and cultures are ultimately 'othered' against and by the British tourist. In both programmes, the dominance, performance and embodiment of British culture, language and beliefs reflect the findings of Andrews (2005, 2011) in the Spanish resort of Magaluf, a site that is spatially dominated by British, white, working-class, heterosexual tourists. This othering of the Spanish and their culture is presented in each programme through a wider dislike and distrust of the Spanish hotel staff and a distaste for Spanish food and culture. Such sentiments have echoes of the campaign to leave[6] the EU and UKIP's political position as held against the EU (UKIP, 2019; Voteleave, 2019). In drawing on the two key Spanish characters in both programmes, it is possible to explore how a distance is created between 'them' (the Spanish and the EU) and 'us' (British tourists and Britain as a whole), with it being possible to argue that the Spanish waiters in both programmes can be theorised as representative of the EU and Britain's relationship to it.

Carlos, the Spanish hotel waiter in *Duty Free*, is presented in a range of negative ways; from his inability to speak good English or understand the correct way to engage with the English (as defined by the English), to his endless clumsiness as represented through him dropping trays of drinks, to fleeing in terror when accosted by the forthright, middle-class Robert. Carlos' reactions and anxiety mirror findings from recent research by the University of Malaga (2018) concerning how tourists' expectations around Spanish hotel staff can be so high that it is likely to create 'conflict and anxiety'. Although Carlos embodies such anxieties upon seeing Robert, for Robert he is conflictive in his interactions with Carlos in a manner that is underpinned through the generalised racist opinions that he holds concerning the Spanish. Robert's opinions echo findings by O'Reilly (2002: 185) in her research with British expatriates in Spain, with one claiming; 'They [the Spanish] are so backward in so many ways... if only they would learn from us [the British]. Some expats have been very important people in their own countries. They could teach these [Spanish] youngsters a thing or two'. Also intersecting with Robert's national identity and his belief concerning the superiority of the British over the Spanish, is his middle-class identity which along with the varied social, cultural and economic capitals it can bestow upon an individual, informs knowledges and ways of 'being and doing' when abroad (Benson, 2013). It is possible to suggest that middle-class Robert carries the confidence in making claims of spatial and cultural dominance over Carlos and his encounters with him. Although Robert is staying in a hotel in Spain, he rules its terraces and Carlos in a manner that has echoes of the British Empire where there are 'still remnants of a superior attitude even while there is an acknowledgement that Europe... needs to be embraced' (O'Reilly, 2002: 184).

Beyond Carlos' exchanges with Robert, he also encounters both David and Linda on several occasions when they furtively meet to continue their clandestine love affair across the three series. Rather than Carlos inhabiting a heroic position as disruptor of the illicit affair (see Smolej, 2010), he is witnessed as a pest, an unwanted distraction to the encounters of David and Linda. This dislike for Carlos and his position as an irritant can be seen through his ability to inadvertently disrupt the freedoms of David and Linda from engaging in their tryst. In a similar way, reflecting on UKIP rhetoric, Tournier-Sol (2015: 143) argues that for those behind the leave campaign in the UK, the EU, just like Carlos, disrupts and curtails the freedoms of the British people and by leaving the EU Britain 'will regain three essential freedoms: freedom of action, freedom of resources and freedom of the people'. When serving drinks or food and knowing 'his place', Carlos is at best tolerated by the British tourists in *Duty Free*, but when he disrupts their freedoms, their British entitlements or forgets 'his place', he is cast as problematic, in a similar way to Britain's wider view concerning the role, usefulness and place of the EU (Fontana & Parsons, 2015).

Mirroring *Duty Free*, the only significant Spanish character in *Benidorm* is the hotel waiter 'Mateo', who for Prieto-Arranz and Casey (2014) can be understood as embodying British representations and imaginings of Spain and the Spanish in several ways. Unlike Carlos, Mateo is initially presented as attractive, sexually able and confident, a man who uses his sexual 'otherness' and good looks to engage in casual sexual encounters with primarily British female tourists (Hughes & Bellis, 2006). However, as the many series of *Benidorm* progress, Mateo is increasingly presented to the viewer as sexually threatening through both the risk of pregnancy or the spread of sexually transmitted infections, and as the British holiday representative Sam claims of Mateo in series six, 'not him, he'll be riddled'. Mateo's perceived sexual risk, and in particular his ability to impregnate British women, echo language used by the leave campaign and the wider Brexit debate concerning 'taking back control of our borders' (Parker & Fleming, 2018) and thus avoiding the diluting of Britishness through the arrival of the immigrant body. Mateo and his uncontrolled sexual exploits are symbolic of the perceived threat and risk of uncontrolled immigration into the UK and the contamination this represents. Such sentiments echo imaginings of Britain during the Victorian Empire, when those nations which made up the Empire were often represented as a distinct other, sexually threatening and savage in comparison to the 'civilised' and 'sexually restrained' homeland (see Bush, 1998).

For a number of those involved in the leave campaign, imaginings concerning the foreign body in terms of its unrestrained sexual and violent threat to 'civilised Britain' or its ability to take on any low paid work and thus deny British citizens a job, were utilised as key moral panics in securing support for growing anti-EU rhetoric. Borrowing from this, it is possible to theorise Mateo and his actions as being representative of other EU nationals' sexual and economic threat to UK citizens. He not only represents a sexual threat to British bodies, both through his ability to seduce British women and with being symbolic of British men's hapless sexual inferiority compared to his sexual prowess, but also his ability to hold down and undertake multiple jobs reflects his economic worth in a neoliberal economy and the potential threat posed by EU workers to British job security. Within the series, Mateo is presented as working in a number of low-skilled jobs such as 'a pool watcher, waiter, security guard, receptionist, porter' (Prieto-Arranz & Casey, 2014: 73), the multiple skills of Mateo and his willingness to take on multiple jobs, can be theorised as representing the perceived threat of large-scale immigration from the EU into the UK for British workers. The economic arguments present in the growing anti-EU stance of political parties and the wider British population, rather than centring on the financial crisis of 2008 and the significant cuts in welfare and social services across the UK, or the long-term structural and economic changes which disadvantage those without university degrees and associated skills in the global marketplace, have had the consequences that many in the UK attribute their economic problems and growing poverty to immigration.

The foreign Spanish body is primarily present in the background in each programme, becoming visible or of worth when services are required by the British tourist at the bar or the reception desk. The low status attributed to Carlos and Mateo as barmen, although a respected occupation in Spain, is witnessed as inferior to even the long-term unemployed Garveys in *Benidorm*. Such images echo existing racist discourses held around the Brexit debate that positions EU immigrants coming into the UK as low skilled and being of only temporary value to the British economy when willing to undertake work that British citizens refuse. For example, recent debates about migration post-Brexit by the Conservative government and the UK Prime Minister Theresa May, has suggested that 'nationals who travel to the UK to work on fruit and vegetable farms will be able to stay for six months before returning' to their home countries (Independent, 2018). In a similar way to Spanish bar staff, EU migrants can then 'never be "us"... The institutionalized and propertized entitlements of "us" require that "they" stand and work as the projected distanced object' (Skeggs, 2004: 171).

An Island Hotel

Within both programmes the perceived threat to British tourists' safety from the Spanish or from 'real Spain' is enacted through a lack of mobility or desire on behalf of the tourists to leave their hotel complexes. For example, in one episode of *Duty Free*, the four main characters visit a 'traditional' Spanish market which is represented as basic, dirty and threatening. Instead of offering an 'authentic' Spanish experience and an enjoyable 'otherness' (see Curtin, 2010), the market is presented as a site in which the British have to avoid being pickpocketed or being overcharged for souvenirs by unscrupulous Spanish stall holders who are presented to the viewer in ways that pertain to subtle signs of 'racism, elitism, and ethno-centrism' (O'Reilly, 2002: 186). This financial distrust of the Spanish is also reflective of Thatcher's five-year war of attrition over the EEC budget, as discussed earlier in this chapter. In another episode involving a boat trip to Morocco, it is clear to the viewer that working-class tourists such as David and Amy do not have the social and economic capital to move freely or confidently beyond the limits of the hotel complex in Spain (see Curtin, 2010; Lawler, 2014). During their return to Spain from Morocco, the safety and security of both David and Amy is put at risk through Amy's perceived naivety and uninformed working-class touristic experiences. Spanish customs officials at the port are positioned as an 'untrusted other', where instead of maintaining the security and safety of the passengers as they pass through the port, the presence of the Spanish customs officials is representative of Spain's problematic and corrupt borders. Such imagery mirrors past and existing debates concerning the inability of the UK to control its own borders while it continues to be a member of the EU, alongside the EU's desire for the borderless movement of its people. As Tournier-Sol (2015: 146) reflects, one of the key and most successful slogans of the leave campaign was for the UK to 'regain control of UK borders'.

In a similar vein, the majority of the working-class guests of the Solana Hotel in *Benidorm* rarely venture beyond its all-inclusive amenities, and when they do, such ventures are often unsuccessful or undertaken with trepidation. In stark contrast to this, the occasional middle-class character in the programme is depicted as making every effort to escape the hotel for the old town of Benidorm for tapas and Spanish 'authenticity', or venture even further afield to escape the sights and sounds of the resort. For Prieto-Arranz and Casey (2014), the middle-class characters in the programme are positioned by the other characters and the viewer alike as not belonging in the resort. For example, in the third series, Janice Garvey asks her husband Mick about another tourist she sees around the pool, 'Who is she? She doesn't belong here, she is far too posh' (Prieto-Arranz & Casey, 2014: 74–75). Her sentiments echo findings by Andrews (2005) in which tourists will 'other' and form disconnections concerning other tourists' social class, bodies and behaviours, deciding if they belong in a particular holiday setting or not. The differences in characters' desire or lack of desire to leave the confines of the hotel and try and access an 'authentic Spain' in *Benidorm* could initially be understood as reflective of the perceived class divide between the urban and wealthy cosmopolitans and the rural or/and urban-poor non-cosmopolitans in the Brexit referendum (see Kwame, 2007). For some commentators, Brexit can then be interpreted as the voice of the angry working class being enacted against both the EU *and* liberal middle-class cosmopolitans in the UK, a view which has gained much public coverage (see Mckenzie, 2016).

However, for Antunucci *et al.* (2017) the profile of Leave voters with respect to education and to socioeconomic conditions is not homogeneous, with Leave voters coming from a range of socioeconomic classes. Instead, it is possible to witness *Benidorm* as reflective of the more complex divide that has underpinned the Brexit debate between those who are pro-Europe/those who wish to leave the hotel and seek out an 'authentic Spain', and those who are anti-Europe/those who only wish to or who can only afford to consume British food, alcohol and media in the confines of the hotel. And as the works of Waldren (1996), King *et al.* (2000) and O'Reilly (2000, 2002) have shown, positive sentiments concerning Europe and personal relationships with Europeans while on holiday or through emigration are present or lacking across all social classes. Instead, the Leave vote may reflect not only the lack of opportunities for particular groups of people and communities across the UK including the working class, but also an increasingly economically vulnerable middle class in an age of austerity (see Antunucci *et al.*, 2017: 214). The 'fear expressed by the Leave voters' (Antunucci *et al.*, 2017) which may be a reaction to new uncertainties in a globalised society is one that mirrors the fear of leaving the Solana Hotel and the unknown uncertainties outside its boundary walls.

Within both *Duty Free* and *Benidorm* it is then possible to approach the hotel as representative of Britain itself. Like Britain, which is an island and separated from mainland Europe by the sea, the hotels in each programme are confined and isolated from the respective resorts of Marbella and Benidorm and their Spanish residents by walls and gates. Joining the EU meant that the UK adopted the free movement of EU citizens across and between all EU nations, eradicating the role of the seas around Britain in limiting the access of the foreign body to the UK. However, with the arrival of Brexit, the UK can once again be detached from the

EU geographically and politically and border controls can be reinstalled (see Tielmann & Schiereck, 2017). But the similarities do not end there; like the UK, each hotel is predominantly populated by white British bodies with each having its own British culture, food, drink, music and entertainment and English being the main language spoken. Although fictional, both *Duty Free* and *Benidorm* in particular, mirror findings in the work of both Andrews (2005, 2011) and Briggs and Turner (2010) on the respective Spanish islands of Mallorca and Ibiza in which hotels and resorts were claimed *to be* British sites through the presence of the Union Jack flag, English media, food and drink, underpinned by drunken and aggressive claims to spatial and cultural dominance by British tourists. Such actions and sentiments echo Nigel Farage's claims of the British that 'we think differently. We behave differently' (cited in Tournier-Sol, 2015: 143), where for the British tourist their difference is marked against the problematic or suspicious otherness of a Spain beyond the confines of the hotel or resort.

Conclusion

Although both *Duty Free* and *Benidorm* were recorded over two decades apart, each share the similarities of being written and aired during complex and difficult economic realities in the UK, alongside the rise of significant anti-EU rhetoric voiced by multiple Conservative governments and the wider right-leaning British media. This chapter does not suggest that either programme was anti-EU in its content, they were made first and foremost to entertain, but it is possible to witness how the many themes present in both programmes captured the social, economic and political malaise of their times.

Duty Free is reflective of a particular historical moment within international tourism of the 1980s, one in which the package holiday is now well established and Spain has developed a place in the British tourists' imagination. But the low-cost flights and large-scale ease of mobility of the 21st century (see Casey, 2010) are yet to arrive for most of the middle and working classes, and in this sense Spain and Europe are still distant, othered, 'exotic' in their portrayal. The familiarity surrounding Spain, or at least the resort of Benidorm that is present in *Benidorm*, reflects the economic and cultural moment in which it first aired in 2007 when the British economy was as strong as it had ever been since the Second World War (Marr, 2007), while also being a moment in which the British population had adapted quickly to low-cost flights and increased mobility across a growingly familiar Europe.

However, in approaching each programme, this chapter has shown that it is possible to witness that both *Duty Free* and *Benidorm* reflect themes to be found in Britain and its politics concerning Europe, the place of the UK in Europe and Brexit itself. The distrust towards the foreign 'other' in each programme is reflective of the wider distrust held by many British politicians, the British press and wider population towards the EU during both the 1980s and the 21st century. Underpinning the seeds of distrust towards the EU as sown by Thatcher in the 1980s, was her own belief in Britain's superiority over her foreign neighbours, with this mirrored in the characters' own superiority and their consequential interactions with Carlos and Mateo in each programme.

During the 1980s, England's industrial north was declining under Thatcher's neoliberal policies and people like David and Amy were losing their jobs, while the south-east of England, the Conservative's heartland, was booming under the rise of the new service economies and the wealth and mobilities they bestowed upon its workers such as Robert and Linda. And although *Benidorm* was initially aired during a time of considerable economic wealth and stability in the UK which had increasingly spread outwards from the south-east of England to some northern cities, the vast majority of *Benidorm* was aired during an intense global recession and the arrival of austerity[7] within the UK, alongside growing political instability, the rise of UKIP and right-wing political groups and growing moral panic concerning the urban poor as represented in the Garveys.

It is possible to theorise that the growing distrust towards the EU and the continued othering of its people, cultures and political ideas, reflects an uncertainty on behalf of the UK of its place within and outside of the EU, and an inability of the UK to address its own failings and the weakening of its worth in a globalised and increasingly fragmented world. *Duty Free* and *Benidorm* present the Spanish and Spain as the humorous and suspicious other, set against the familiar British tourists and their ways of performing and embodying Britishness, with both programmes reflecting key historical moments concerning a growing fear, discomfort and increased racism towards those both inside and outside Britain's borders. In reading the programmes, it is possible to witness that it is the British who sit uncomfortably within Europe through their unwillingness to change and move beyond both their island and Victorian Empire mentality. And in a similar vein, the inevitable return of the characters in *Duty Free* and *Benidorm* to the UK at the end of their holidays mirrors that Brexit and Britain's withdrawal from the EU is almost an inevitable outcome from its long and uncomfortable relationship with itself and the EU.

Notes

(1) Commonly referred to as 'Thatcher'.
(2) The EEC stands for European Economic Community.
(3) The Labour Party is one of two main political parties in the UK and is understood as a left-leaning and liberal party. The Conservative Party is the other main political party in the UK and is right leaning and conservative in its values.

(4) In both *Duty Free* and *Benidorm*, the class of characters can be witnessed in the different ways that each dresses and presents themselves. For example, in *Duty Free*, middle-class Robert is often wearing a lightweight summer suit and Panama hat, not dissimilar to those worn by middle- and upper-class British colonisers in countries such as India during the age of the British Empire. In contrast, David's character only has one 'good suit' that he wears in Marbella to try and appear to 'be' middle class in appearance, even if his suit is not suitable for the hot Spanish climate. His lack of clothing choice and his inability to afford or wear the 'correct' clothing indicates to the viewer that he is working class.
(5) The 'chav class' can be understood as a growing group of people in the UK who are characterised as being long-term unemployed, poorly educated, live in deprived areas of British cities and are socially, spatially and culturally excluded from 'mainstream' society. An excellent discussion of this group can be found in the work of Jones (2011).
(6) Commonly referred to as the 'leave campaign' in the UK.
(7) Austerity refers to the tough and difficult economic measures adopted by successive Conservative governments since the recession of 2008. Such measures have included staggering reductions and cutbacks of public services across the UK.

References

Allen, K. and Mendick, H. (2012) Keeping it real? Social class, young people and 'authenticity' in reality TV. *Sociology* 47 (3), 460–476.
Andrews, H. (2005) Feeling at home: Embodying Britishness in a Spanish charter tourist resort. *Tourist Studies* 5 (3), 247–266.
Andrews, H. (2011) *The British on Holiday: Charter Tourism, Identity and Consumption.* Bristol: Channel View Publications.
Antunucci, L., Horvath, L., Kutiyski, Y. and Krouwel, A. (2017) The malaise of the squeezed middle class: Challenging the narrative of the left behind Brexiter. *Competition & Change* 21 (3), 211–229.
Barker, C. (1999) *Television, Globalization and Cultural Identities.* Maidenhead: Open University Press/McGraw-Hill.
Benson, M. (2013) Living the 'real' dream in la France Profonde? Lifestyle migration, social distinction, and the authentic of everyday life. *Anthropological Quarterly* 86 (2), 501–525.
Bote, G.V. and Sinclair, M.T. (1996) Tourism demand and supply in Spain. In M. Barke, J. Towner and M.T. Newton (eds) *Tourism in Spain: Critical Issues* (pp. 65–88). Wallingford: CABI.
Bray, R. (2001) The 1950s. In R. Bray and V. Ratiz (eds) *Flight to the Sun* (pp. 34–54). London: Continuum.
Briggs, D. and Turner, T. (2010) Understanding British youth behaviours on holiday in Ibiza. *International Journal of Culture, Tourism and Hospitality Research* 6 (1), 81–90.
Brown, D. (1984) Thatcher settles for 66pc rebate. *The Guardian*, 27 June. See https://www.theguardian.com/politics/1984/jun/27/past.eu (accessed 13 March 2019).
Bush, J. (1998) Edwardian ladies and the 'race' dimension of the British imperialism. *Women's Studies International Forum* 21 (3), 277–289.
Casey, M. (2010) Low cost air travel: Welcome aboard? *Tourist Studies* 10 (2), 175–191.
Curtin, S. (2010) What makes for memorable wildlife encounters? Revelations from 'serious' wildlife tourists. *Journal of Ecotourism* 9, 149–168.
Fontana, C. and Parsons, G. (2015) 'One woman's prejudice': Did Margaret Thatcher cause Britain's anti-Europeanism? *Journal of Common Market Studies* 53 (1), 89–105.
Hall, S. (1980) Encoding/decoding. In S. Hall and D. Hobson (eds) *Culture, Media, Language* (pp. 128–138). London: Routledge.

Hughes, K. and Bellis, M.A. (2006) Sexual behaviour among casual workers in an international nightlife resort: A case control study. *BMC Public Health* 6 (39), 1–5. doi:10.1186/1471-2458-6

Independent, the (2018) Brexit: Farmers allowed to recruit 2,500 migrants a year under new government plan to plug seasonal workforce gap. See https://www.independent.co.uk/news/uk/home-news/brexit-farmers-migrants-seasonal-workers-a8524871.html (accessed 12 January 2019).

Jenson, S. (2012) Othering, identity formation, and agency. *Qualitative Studies* 2 (2), 63–76.

Jones, O. (2011) *Chavs: The Demonization of the Working Class*. London: Verso.

King, R., Warnes, T. and Williams, A. (2000) *Sunset Lives; British Retirement Migration to the Mediterranean*. Oxford; Berg Publishers.

Kwame, A.A. (2007) *Cosmopolitanism: Ethics in a World of Strangers*. London: Penguin.

Lawler, S. (2014) *Identity: Sociological Perspectives*. London: Polity.

Lewis, J. (2004) The meaning of real life. In L. Ouellette and S. Murray (eds) *Reality TV: Remaking Television Culture* (pp. 288–303). London: New York University Press.

Lipton, M. (2008) Principally boys? Gender dynamics and casting practices in modern British pantomime. *Contemporary Theatre Review* 18 (4), 470–486.

MacLure, M. (2003) *Discourses in Educational and Social Research*. Buckingham: Open University Press.

Marr, A. (2007) *Andrew Marrs History of Modern Britain* (Episode 5). BBC4.

Mckenzie, L. (2016) Brexit is the only way the working class can change anything. Opinion. *The Guardian*. See https://www.theguardian.com/commentisfree/2016/jun/15/brexit-working-class-sick-racist-eu-referendum (accessed 1 July 2020).

Nayak, A. (2006) Displaced masculinities: Chavs, youth and class in the post-industrial city. *Sociology* 40 (5), 813–831.

O'Reilly, K. (2000) *The British on the Costa del Sol*. London: Routledge.

O'Reilly, K. (2002) Britain in Europe/the British in Spain; Exploring Britain's changing relationship to the Other through the attitudes of its emigrants. *Nations and Nationalities* 8 (2), 179–193.

Parker, G. and Fleming, S. (2014) David Cameron: Divided loyalties. *Financial Times*, 29 January.

Pickard, J. and Gordon, S. (2018) Theresa May says Brexit deal will give UK 'control over our borders'. *Financial Times*. See https://www.ft.com/content/dccd97a4-ebed-11e8-89c8-d36339d835c0 (accessed 14 May 2019).

Prieto-Arranz, J.I. and Casey, M. (2014) The British working class on holiday: A critical reading of ITV's Benidorm. *Journal of Tourism and Cultural Change* 12 (1), 68–83.

Schroder, K. (2007) Media discourse analysis: Researching cultural meanings from inception to reception. *Textual Cultures: Texts, Contexts, Interpretations* 2 (2), 77–99.

Skeggs, B. (2004) *Class, Self, Culture*. London: Routledge.

Smolej, M. (2010) Constructing ideal victims? Violence narratives in Finnish crime-appeal programming. *Crime, Media, Culture: An International Journal* 6 (1), 69–85.

Spencer, S. (2011) *Visual Research Methods in the Social Sciences*. London: Taylor & Francis.

Tielmann, A. and Schiereck, D. (2017) Arising borders and the value of logistic companies: Evidence from the Brexit referendum in Great Britain. *Finance Research Letters* 20, 22–28.

Tournier-Sol, K. (2015) Reworking the Eurosceptic and Conservative traditions into a populist narrative: UKIP's winning formula? *Journal of Common Market Studies* 53 (1), 140–156.

Turner, L. and Ash, J. (1975) *The Golden Hordes: International Tourism and the Pleasure Periphery*. London: Constable.

Valenzuela, M. (1988) Spain: The phenomenon of mass tourism. In A. Williams and G. Shaw (eds) *Tourism and Economic Development* (pp. 40–60). London: Wiley.

UKIP (2019) See https://www.ukip.org/ukip-page.php?id=05 (accessed 24 March 2019).
University of Malaga (2018) Management strategies help reduce emotional exhaustion of team members in Spanish hotels and restaurants. *Human Resource Management International Digest* 26 (5), 16–17.
Voteleave (2019) See http://www.voteleavetakecontrol.org/our_case.html (accessed 24 March 2019).
Waldren, J. (1996) *Insiders and Outsiders: Paradise and Reality in Mallorca*. Oxford: Berghahn Books.

8 Taking Back Control: The Freedom of the Holiday

Hazel Andrews

Introduction

> Ask an Englishman what nation in the world enjoys most freedom, and he immediately answers, his own. Ask him in what that freedom principally consists, and he is instantly silent[1]
>
> Oliver Goldsmith in Franklin (1989)

Like most plebiscites, the referendum on the UK's membership of the European Union (EU) on 23 April 2016 was preceded by a campaign to persuade voters to vote one way or the other. In the case of the EU referendum, the campaigners sought to promote either 'yes' to remain in the EU or 'no' to leave. Unlike campaigns for electoral representatives in the UK, the choice was an either/or which saw a unity of sorts in that members of the political divide were united based on vote yes to stay or vote no to leave. There were two official campaign groups, Vote Leave and Britain Stronger in Europe, their respective names speaking obviously for which side of the divide they fell. Vote Leave was headed by the Conservative Party MP Boris Johnson (and now, at the time of writing, the UK's prime minister) and included among its members the then Labour Party MP Gisela Stuart. The Britain Stronger in Europe campaign included the then Conservative Party leader and prime minister David Cameron and former Labour Party leader and prime minister Tony Blair. The overall point being that whether to leave or remain was not based on which side of the 'traditional' political party lines one normally sat.

A central tenet of Vote Leave was 'take back control'.[2] This appealed to ideas that in being a member of the EU, the UK had somehow lost something, principally sovereignty. Indeed, in May 2013, the British daily tabloid newspaper *The Sun* ran with a strap line that said 'Save our Country' printed against a backdrop of the Union Jack. The headline features a depiction of Queen Elizabeth I with the caption above reading '1588. We saw off the Spanish', next to this is a picture of Admiral Horatio Nelson with the wording above declaring '1805. We saw off the French' and alongside this is an image of Winston Churchill below the caption '1940. We saw off the Germans'. Then, beneath what appears to be a tear in the paper is a photo of Tony Blair next to the caption '2003 Blair surrenders Britain to Europe'.[3] And, if one follows the story to the inside of the newspaper, *The Sun*'s political editor Trevor Kavanagh claims that in what is 'The biggest betrayal of our history', Tony Blair is on the verge of 'sign[ing] away 1,000 years of British sovereignty' (p. 6) in reference to agreeing to the terms of a new EU constitution, which, it was argued, was a move that took powers away from the UK government and gave them to the EU. Sensationalist headlines and warnings of threats to British sovereignty are not unusual in the UK tabloid press. Indeed, in October 2003, *The Daily Express* claimed that there was an 'EU plot to rename Trafalgar Square and Waterloo Station'. In March 2006, *The Sunday Express* newspaper ran a story alleging that there was a 'Brussels plot to wipe Britain off the map'. I could go on; the list of similar stories and headlines is endless. Their appearance has continued post-referendum with the focus often targeted at those who are seen to be attempting to thwart the UK's departure from the EU, the most infamous of these headlines is arguably the tabloid *Daily Mail*'s 'Enemies of the People'[4] front page reacting to the high court judgement in the case of *R (Miller) v Secretary of State* for Exiting the European Union[5] that Parliament must be able to vote on the Brexit process. An irony here is that in 'taking back control' by voting to leave the EU, as claimed by Vote Leave, control in the form of parliamentary sovereignty – until the Miller intervention – was, in effect, being denied. A counterclaim by Vote Leave, as exemplified in the *Daily Mail* headlines, was that the 'will of the people' was being obstructed.

Doubtless, the debates about Parliament's role will continue. What piques my interest here, however, is the word 'control'. What does it mean to have control, be out of or in control? Looking at some of the campaign material for Vote Leave, control appears to be related to being in command of events and situations, for example, that leaving the EU would mean 'We'll be in charge of our own borders' and that without Brexit 'Immigration will continue to be out of control'. However, upon departure from the EU, 'We'll be free to trade with the whole world… we'll be free to seize new opportunities'.[6] To be in control then means to be free.

Such rhetoric has continued in the political wrangling that has characterised the debate in the years since the referendum. As the deadline for departure from the EU came and went in March 2019 with the UK still a member, the country was required to take part in the European elections of May 2019 for members of the European Parliament (MEPs). Apparent frustrations with existing MPs led to the launch of a new political party in April 2019. Founded by the one-time leader of UKIP Nigel Farage, the Brexit Party attracted support from those new, and not so new, to the world of politics. One such candidate was former Conservative Party MP Ann Widdecombe, who in the EU elections was elected as an MEP for south-west England. In her maiden speech and as the first member of the Brexit Party to speak to the EU Parliament, Widdecombe described the UK's proposed departure from the EU as an emancipation from slavery. She argued 'There is a pattern consistent throughout history of oppressed people turning on their oppressors, slaves against their owners, the peasantry against the feudal barons, colonies against empires, and that is why Britain is leaving ... It doesn't matter which language you use; we are leaving and we are pleased to be going. *Nous allons* [sic], *wir gehen*, we are off!'[7] Here again the justification for leaving the EU and 'taking back control' is couched in the language of freedom. And, as Kelty (2017: 165) notes, freedom is often 'employed by elites and the impoverished alike, taking shape as a guiding principle, justifying action as much as providing an incentive, a goal or a reason'.

In the acknowledgements section of their edited collection *Freedom in Practice. Governance, Autonomy and Liberty in the Everyday*, Moises Lino e Silva and Huon Wardle (2017: ix) note that '"Freedom" is one [of] the most fiercely contested words in contemporary global experience'. In tracing the lineage of anthropological discussion on freedom, they attest that it is not a word that has often been considered by the discipline, pointing to Malinowski's (1947) observation that trying to find a definition of what it means in everyday life leads to 'semantic chaos'. Another layer of complexity is added in that, as Kelty (2017: 164) notes, different perspectives are brought by different disciplinary analyses of the notion. Further, as a concept it is not one easily identifiable in or translatable between other cultures. Indeed, Hideko Mitsui (2017: 54) points out that in Japan the concept of freedom was only introduced in the early 20th century.

Lino e Silva and Wardle argue that freedom belongs to a group of words that includes liberty, self-determination and autonomy (although all have a different etymology with the latter being more commonly used in anthropology). It is, they claim, hard to disentangle ideas of freedom from Judeo-Christian ideas of 'free will' and, when compared with autonomy, freedom becomes an altogether more ill-defined term. In addition, although freedom may be used interchangeably with liberty and autonomy, there are subtle differences. As Nigel Rapport (2017) argues, liberty has been a more favoured word in the English language than freedom given its roots in Latin that linked it, for the Victorians, with Imperial Rome and the idea of empire. By contrast, freedom has Anglo-Saxon origins. In his examination of liberty and freedom through the writings of English artist Stanley Spencer, Rapport argues that liberty is given by others whereas freedom refers more to an ontological state of being. He states, 'Freedom I have defined as a state of being, part of the condition of organic (including human) life. Freedom describes the way in which life is in itself' (Rapport, 2017: 51). This is a view similar to that of Michael Jackson (2013: 127), who attests that freedom is one of those words (of which others are dignity, respect, honour) that has existential value but defies definition.

Jackson (2013: 247–248, emphasis in original) goes on to suggest 'that we have to give up on the idea that we can substantivize words such as… freedom…, and learn to use these terms as ways of capturing *aspects* of what is at play in any human situation'. This idea fits well with Lino e Silva and Wardle's (2017: 30) assertion that anthropological perspectives on freedom should not focus so much on trying to pin down a definition of freedom but rather should seek to understand how people think about and live freedom for themselves. In heeding this advice, this chapter is an exploration of freedom, and by corollary control, as they manifest themselves in touristic practices in the Spanish charter holiday resorts of Magaluf and Palmanova on the Mediterranean island of Mallorca. Based on ethnographic fieldwork involving periods of participant observation which identified an 'effervescent Britishness' (Andrews, 2011) enacted by tourists and encouraged by many of the mediators of tourists' experiences, the chapter will illuminate what control and freedom both mean in this setting and thus contribute to anthropological debates about freedom. I'll begin with an overview of the research setting.

Effervescent Britishness in Mallorca

Mallorca, the largest of the Balearic Islands, has been attracting international visitors for leisure purposes since at least the 19th century. With the development of tourism products in the 1950s alongside the viewing of tourism development as a way to earn foreign income and create jobs, the island was identified by the Franco government as one among many places in Spain suitable for investment in terms of tourism (Andrews, 2017a). Many resorts grew in response to the increasing demands of a wealthier and 'spare time' richer population in Northern Europe seeking time away from the demands of everyday life in a setting providing guaranteed good weather. Palmanova and Magaluf are two such resorts. Situated south-west of Mallorca's capital Palma on the Bay of Palma in the local municipality of Calvià, they have enjoyed great success in terms of tourist numbers and profits. This has been to such an extent that Calvià has been one of the richest local authorities in Spain and Europe (Andrews, 2017a). Like other of the tourism destinations on the island, Palmanova and Magaluf have grown with an appeal distinct to particular nationalities. Thus, some places became known for being German, for example, S'Arenal (see Storch and Nassenstein, this volume) on the opposite side of the Bay of Palma from Magaluf and Palmanova has its 'Schinkenstrasse' (Ham Street) and 'Bierstrasse' (Beer Street) that appeal to a predominately German clientele many of whom are seeking the excesses of a party-style holiday involving heavy drinking in a setting that offers some 'home comforts' with plenty of wurst (sausages), the familiarity of the German language and entertainers from Germany. Magaluf and Palmanova, by contrast, are decidedly British, with Punta Ballena – the main drag of Magaluf – given over to catering for predominately British tourists in nightclubs and cafés that are tailored to British tastes.

It is not that either so-called German resorts or British resorts are exclusive to either nationality, but rather it is where there tends to be a concentration of people hailing from these particular countries and identifying themselves in a particular way, which in the case of Magaluf and Palmanova is British. I fully recognise the complexities of Britishness (I have explored this elsewhere in relation to the British abroad [see, for example, Andrews, 2011, 2017b; and also O'Reilly, 2000]), and its subsuming of other ethnic descriptors in the UK (Scottish, English, Welsh), but without being able to ascertain from exactly which part of the UK someone derived, it is necessary to use a term that is generally understood to refer to the citizens of the UK.

I make my claim for the Britishness of Palmanova and Magaluf based on the dominance of tourists from the UK, the ubiquitous use of the English language, including the availability of various forms of English language media – including *The Sun*, *Daily Express* and *Daily Mail* newspapers and UK-based TV shows featuring Saturday night entertainment, sporting fixtures and soap operas. In addition, food, beverages and entertainers are all imported from the UK to serve tourists' tastes. Thus, it is possible to eat, for example, breakfast cereals, bread, bacon and sausages all from the UK and to drink British milk and buy alcohol sold as an imperial pint, even being able to, in some cases, pay in pounds sterling. In addition, the numerous establishments catering for tourists outside of the hotels were often named in a familiar British fashion – The Red Lion, The White Horse as examples. This was particularly so in the case of Magaluf, where the concentration of party, 18–30-style holiday-making that has earned the resort the nickname of 'Shagaluf'[8] took place. Although Palmanova was also frequented by a mainly British clientele, the tone of the resort was different with fewer obvious references in shop and café names to Britishness and a generally calmer atmosphere at night. The Union Jack flag was no less conspicuous, however, advertising British consumables based on 'cooking like mum's' alongside the aforementioned signals of Britishness running through both resorts.

I first went to Mallorca in 1997 and followed this initial week-long visit with longer stays in 1998 and 1999. I next visited in 2009, 2015 and, most recently, 2018. The bulk of my data were collected in the earlier visits, with my return in 2015 confirming that, at least in terms of what appeared to be on offer, little had changed. The out-of-season visit during 2018 identified that there were signs welcoming stag and hen parties into several bars, a trend that had not previously been as prominent. During the earliest visits, I engaged with many touristic activities

– sunbathing, sightseeing, souvenir buying, eating out, watching hotel and other night-time entertainment and going on bar crawls. I spoke to tourists and members of the expatriate community (many of whom worked in tourism-related businesses serving the British on holiday) about their holidaymaking practices, experiences of living in Mallorca and life in the UK. Among the tourists were first-time visitors and those tourists who had visited Mallorca many times, staying in the same resort, in the same hotel and sometimes seeking out the same room, others owned holiday homes. The British came from all corners of the UK, but most notably from what might be described as The North (see Andrews, 2017b); they were also predominately white and working class defined as being engaged in mainly unskilled or semi-skilled employment. All names used are pseudonyms.

A theme to emerge from my engagements with tourists and members of the expatriate community was a feeling that a holiday or lifestyle move to either Magaluf or Palmanova was not necessarily about a search for difference or for engagement with another culture that tourism is often purported to be about. For example, Dave, a tourist from Dundee, explains 'I like somewhere commercial. I've been to Tenerife and Torremolinos …they are all the same really. When you come to Spain you know what you are getting. I like to come to a place where I know I'll be able to get a fried breakfast'. He said he would not like to visit India as for him the country was an example of somewhere that was not developed enough for his tastes. For many of the tourists then it was rather a sense of the familiarity that attracted, feelings that it was possible to be more British and demonstrate Britishness in the resorts compared to being at home in the UK. There was a feeling that the UK had somehow changed, lost something and was under threat from outside forces represented by various forms of inward migration to the country, as well as advanced and continued globalisation processes and the EU. As one older male tourist who owned a holiday flat advised, 'England isn't England anymore'. In another discussion about tourists' motivations with a younger couple – Jill and Craig – from Yorkshire on their honeymoon, the UK's relationship with the EU was highlighted. For Jill the answer to reasons for taking a holiday to Mallorca was the weather; however, Craig stated, 'it's because of British bureaucracy'. He argued that 'we're supposed to be a member of the EEC [sic], but we're not governed by the same rules… we can't drink all day'. He went on to comment on the price differences between the UK and Spain for a bottle of whiskey, petrol and cigarettes in which he claimed that as all these commodities were cheaper in Spain, the UK government was 'screwing the taxpayers'. He observed that 'if we're a member of Europe the prices should be equivalent all the way through'. For Craig, citizens of the other European member states had an easier, freer life compared to those in the UK. He cited what he claimed was new health and safety legislation and road tax as other areas of inequality, arguing that taxes in the UK should not be so high and if the situation continued he would move to Spain, claiming that UK bureaucracy undermined the idea of a 'Great' Britain because 'we live in an oppressive society'.

The idea that the UK has changed both internally and in terms of its position on the world stage, for which membership of the EU can, in part, be blamed, is a well-rehearsed observation. Indeed, Lunn (1996: 84) notes that '[g]reat concern is frequently expressed about the loss of British sovereignty which seems implicit in a greater acceptance of European integration'. The situation has been further exacerbated by the demise of the British Empire which reduced the UK's reach and influence as a global authority. Being a member of the EU has, for some, compounded the sense of loss as further integration policies between the member states have been understood to threaten the independence and coherence of the nation-state, as the newspaper headlines discussed earlier indicate. As Cesarani (1996: 64) observes, this has been an ongoing process since the end of the Second World War after which 'National identity, nationality and citizenship were increasingly inflected by global patterns of migration and movements of the international economy, notably Britain's relative decline and the formation of the European Economic Community'. The loss of empire also signals a loss of purpose in the world adding to the insecurities about the outward facing image of a country that needs to reposition itself in global affairs from a stance that is no longer supported by having power over territory and people. This has caused some to suggest that the UK suffers from a lack of self-confidence and feelings of impotence (see, for example, Cinnirella, 2000; Triandafyllidou, 2002).

The sense of insecurity in the face of a foreign other on an individual level was exploited by some of the mediators of the tourism experience, particularly the commission-hungry tour operator representatives (reps), who cautioned against the dangers of too much engagement with locals; for example, tourists were advised not to use local buses or to venture to other parts of the island outside of an organised tour. 'Us' versus 'Them' was a theme that also emerged in a wide variety of entertainment, most notably the night-time entertainment Pirates Adventure which symbolically enacts conflicts with France, Spain and Germany (Andrews, 2009, 2014); but which is also manifest in other shows that highlight not only the differences between the UK and other parts of Europe, but also differences within the UK itself (Andrews, 2017b). Thus, a sense of self relating to gendered, regional and most markedly national identity was promoted, celebrated, practiced and sought out. And all of this was within a context predicated on ideas of freedom and losing control. Freedom on holiday was not linked to recovering a sense of control but was couched in terms of losing control. This is unlike the Vote Leave campaign for Brexit in which taking back control is a necessary part of gaining freedom.

On Holiday with the Free

The sentiments expressed by Craig about the UK having an oppressive society would seem to chime well with cries for freedom championed by Leave UK and Ann Widdicombe's speech proclaiming departure from the EU would lead to emancipation of some sort for the populace. However, according to Craig's words, it appears that the control and thus lack of freedom he felt was the fault of the UK government rather than the EU and that this was a driver for holiday-taking practices. Indeed, another male tourist, who originally hailed from the UK, but now resided in Jamaica explained his decision to relocate to Jamaica based on, in part, the greater sense of freedom that he felt his new home offered him 'to live life how I want to live it', although he missed aspects of his former life in relation to culture as manifest in 'the theatre, entertainment, beer, bacon and sausages'. Another expat, who lived and worked in Magaluf, was an economic migrant, her husband having lost his job 'when Maggie Thatcher closed down the coal mines'. For Brenda, being away from her UK Yorkshire hometown equated with the freedom to go out and about and not to conform to the expectation of staying at home to cook the Sunday dinner while her husband was in the pub. The words 'to live life how I want to live it' and the practise of 'coming and going as I want' accord with Rapport's (2017: 51) definition that freedom is 'a state of being... the way in which life is in itself'. The words relating to freedom spoken by the tourists are freedom 'to do' and the freedom 'from' – the latter a form of the negative freedom identified by Isiah Berlin (1990, in Rapport, 2017: 35) which becomes part of the embodied and experiential aspects of being on holiday.

Embodying freedom

The idea that freedom was to be found on holiday rather than at home was echoed by other tourists and was embodied in their practices. For example, Tony, a tourist in his twenties from Wolverhampton, holidaying with his girlfriend was, like other tourists to Magaluf, in part attracted by the nightlife the resort offers. Upon asking the couple what the attractions of clubbing in Magaluf were compared to the UK, Tony replies: 'you couldn't go to a nightclub in t-shirt and shorts, they are too strict in England, here there are no restrictions'. Another tourist, Sheila from Nottingham on holiday with her husband and two teenage sons, had not been aware of the nightlife culture that Magaluf offered, nor its appeal to ideas of Britishness prior to arriving in the resort. Although she and her family had been disappointed with what they saw as a lack of Spanish culture – unusually for tourists to Magaluf they were interested in the foreign other – she had nevertheless enjoyed her holiday saying, 'I've had a real sense of freedom... I don't know if it's because we both have demanding jobs. I haven't missed doing the dishes and cooking'.

For her, freedom was a release from the requirements of the quotidian world of going to work and attending to domestic chores. For another female tourist, Emma, ideas of freedom related mainly to the holiday experience of her 11-year-old daughter. Emma explained that she and her husband referred to Mallorca as 'fantasy island'. They were regular visitors to Palmanova enjoying the warmth and sunshine, and what Emma described as a 'sense of freedom about the area'. Although on the occasion of my conversation with Emma, she and her husband were travelling without their daughter, Emma explained that when she did accompany them, 'there is so much freedom for her she can go up and down the beach and we know it and we don't need to worry'. For Emma, freedom equated to safety for her daughter. The freedom from danger that Emma expresses along with Brenda's and Sheila's freedom from domestic work are further examples of Berlin's ideas of negative freedom allowing 'the freedom from all manner of debilitating conditions that might act as hindrance to the individual citizen fulfilling himself or herself' (Rapport, 2017: 35).

In her reference to Mallorca as 'fantasy island', Emma is alluding to the myths woven around holiday destinations in their marketing that allow potential tourists to imagine themselves in a particular place, stimulating expectations that, although open to disappointment, are also responsible for feeding the practice of being on holiday. Thus, the opportunity to eat and drink (and have sex) as much as one likes is a recurrent theme in the marketing of tourism destinations that fuels notions that touristic practice is about doing what one wants.

The idea of doing what one wants is embraced by many tourists and reports of 'bad behaviour' – hotel and apartment rooms damaged, excessive drinking, making too much noise, exposing too much flesh, pushing to the front in the hotel entertainment – all garnered criticism from fellow tourists, local people and the tour operator reps who were often the source of facilitating activities specifically designed to encourage one to 'lose control'. Nevertheless, the culprits of 'misbehaviour' were seen as 'animals' or behaved like 'monkeys'. As Martin, a British student working in a café in Palmanova for the summer of 1998 advises, 'tourism is horrible… in Magaluf [where he had worked during a previous summer] you get to see the other side of people. Because they're in a foreign country they think they can do what they want'.

Herein lies the rub: freedom and control sit alongside each other as tourists are sold ideas of freedom which on delivery in practice are within the bounds of tour operator control and expected codes of behaviour, some of which some tourists bring with them. Monica Hanefors (2001) argues that tourists leave their 'ordinary' home environments to experience life in a 'non-ordinary' destination which takes place within a 'tourist bubble'. She claims that tourists anticipate and expect what life in the tourist bubble will be like, adding that the location of the 'bubble' away from home implies freedom, but that the tourists add to it their own structure, discipline and social control.

Controlling freedom

> If I go into someone else's hotel I sit at the back, but they don't... he was climbing over the chairs to get to the front.

This statement was made by a male retiree from South Wales on holiday in Palmanova with his wife and granddaughter. It is from part of a conversation in which he was explaining how the pool-side sunbeds were reserved from early morning, meaning that he and his family had generally been to the same spot on the beach every day. He said that he had found the other people staying in the hotel to be cliquey and that their 'sticking together' was particularly noticeable during the evening entertainment when people not staying in the hotel would also attend and that these outsiders did not appear to know how to behave. As a result of the 'bad' behaviour of the other tourists, he and his family had tended to visit the café-bars in the vicinity of the hotel for their evening entertainment. For this tourist, there were clear protocols associated with behaviour relating to what rights people had to behave in a certain way if they had not purchased it. There was a sense of entitlement from having bought something (a hotel room, a package) which was being undermined or controlled by the behaviour of others exercising their holiday freedom. This tourist was in effect pushed out from the hotel environs. Arguably, he did not feel a sense of freedom on holiday; his decisions on where to sunbathe and spend the evening being in effect controlled by the behaviour of others who were unaware of, or insensitive to, the injustice of reserving sunbeds or sitting in the front row for entertainment. Here, individual freedom was at odds with the freedom of others. As Rapport (2017: 35) observes, freedom appears more about the individual rather than interpersonal relationships 'irrespective of and independent of social context'. The tourist climbing over the seats to get to the front row before the performance started was concerned with himself, appearing unconcerned that he was operating within the context of 'someone else's hotel'.

As noted, one of the roles of tour operator representatives is to mediate as much as possible the tourists' experiences. This comes in various guises of organising activities including excursions to other parts of the island for sightseeing or shopping. One male tourist, on holiday with his family in Palmanova, had been on a trip to the Caves of Drach (a very popular attraction on the East Coast). The holiday is his first abroad. He tells me how much he and his family have enjoyed the vacation, advising that he would like to return the following year, hire a motorbike or car and explore the island more. In relation to the Caves of Drach, he advises, 'I would have liked to drive myself and be outside the confines of the coach, not be restricted by it'. During our conversation, he also explains that he had read a book about someone who had travelled to India, bought an Enfield bike and ridden it back to the UK across India and Europe. He says that he would like to travel around India on such a bike, explaining that he has been biking since he was eight, 'it's in your blood, the freedom of it... When life gets me down me and my son go off on the bike'. The coach trip was a restricting experience for him not just in the confines of the coach but also in the enclosed nature of the attraction itself – underground caves that other tourists described as 'claustrophobic', 'stuffy and unable to breath in'. The bike represents freedom of movement and self-autonomy.

Similar feelings of freedom through movement were expressed by a tourist following an organised trip to Andratx market and Andratx port. Situated south-west of Magaluf and Palmanova, the town of Andratx is home to a weekly market. It is a network of densely packed streets selling a range of goods including food, household items, clothes, bags, watches, etc. It is one of the main excursions offered by tour operators as well as independent travel companies. During the summer months, it gets very crowded as coach loads of tourists visit, eager for a bargain. Although taking place in the open, there is a feeling of claustrophobia as the volume of tourists combined with the numerous stalls all with their canopies blocking out the sunlight and trapping in the heat contribute to feelings of enclosure, as one male tourist commented, 'it was claustrophobic'. For another tourist, Vicky from north-east England holidaying with her family in Palmanova the market was 'a bit too enclosed'. She had preferred the visit to the port (after a couple of hours at the market, the excursion then takes the tourists to the Port of Andratx which is home to a marina full of yachts, including, in 1998, one reportedly owned by Peter Stringfellow[9]). The tourists get to see a contrast in lifestyle – the bargains to be had at the market compared to the expense of a luxury yacht. For Vicky, the yachts were representative of a lifestyle she aspired to, saying 'the port is beautiful. I've got my eye on one of the apartments overlooking the harbour for sale. There were some nice yachts in the harbour. Our [hotel] room overlooks the sea... there are some nice yachts outside. I'd like that kind of holiday, a few days here and decide you've had enough and go somewhere else'.

The 'taking back control of our borders' promoted by the Vote Leave campaign is not only about inhibiting the movement of people into the UK, but, by corollary, is also about constraining movement from the UK, whether this be through such actual processes as visa requirements, increased travel insurance or subtle changes in psychological dispositions towards the world out there. Such a desire to control and contain seems to be at odds with the sentiments expressed by those who find freedom through travel. It is interesting to note at this juncture, Buzard's (1993: 19) observation about travel from the UK to continental Europe following the end of the Napoleonic Wars: 'By effectively closing the Continent to British travellers for twenty years, the Napoleonic Wars both frustrated and nourished a demand for foreign travel; after 1815 Britons seemed to explode across the Channel, heading abroad in greater numbers than ever before'.

Conclusion: The Free are Always Freer Over There

The idea that life is better, or the grass is greener, elsewhere appears as a familiar spatially infused trope across cultures and through time. It is often couched in terms of discovering an elsewhere in which we would be free from the woes of everyday life, all desires would be satisfied and we would live in peace and harmony with each other (for examples, see Andrews, 2011: 6). According to St Francis, 'all men [sic] are in some part of their being *homines viatores…* perennial movers' (Pagden, 1993: 2). Doubtless, some of us are more desirous of moving than others; nevertheless, millions do move every year, travelling for their holidays, and people from the UK are part of these migrations. What underlies the motivation for being a tourist is as complex and varied as those who engage in the practice; nevertheless, there are associations with ideas of freedom. By-and-large this is a negative freedom in which individuals feel unconstrained by life at home whether that be chores of everyday living or a sense of not having to conform to certain regulations around drinking alcohol or what clothes to wear, as examples, or be under the rule, in the words of Craig, of an oppressive government.

Both sides of the Brexit debate have had recourse to the language of freedom, perhaps most evident in Vote Leave's claims that 'taking back control' is the basis for being free. The irony is that for those holidaying in Magaluf and Palmanova, too much control is the antithesis of the freedom they seek in touristic practices. The practice of freedom on holiday (regardless of the reality of being able to lose control or do as one likes) is under threat in a post-Brexit Britain that has 'taken back control'; because in that process of apparently reasserting authority over the UK's borders those British people who wish to holiday in the EU will most likely face border controls and restrictions to their own freedom of movement that will impinge on their holidaymaking practices.

This chapter opened with Oliver Goldsmith's words, cited in Jill Franklin's (1989) essay *The Liberty of the Park*. While still wishing to acknowledge the subtle differences between ideas of liberty and freedom that have been outlined in the preceding discussion, it seems pertinent to quote some of Franklin's own concluding words:

> According to the myth, the park is a place expressive of English liberty, a natural landscape of freely-laid-out grass, water and trees. Ugly scenes of hard work and views of unsightly villages have been banished, though

peasants may loll and cowherds rest.... Outside in the moral domain, the idle waste has been cleared of villains and poachers and instead the law-abiding husbandman toils in manly submission in the beauteous order of the rich, square fields. (Franklin, 1989: 155)

Vote Leave campaigned to 'take back control' and assert freedom for the British but at the expense of freedom of movement not just for those (read the villains and poachers of Franklin's words) wishing to enter the UK, but also, by corollary, for those wishing to leave. There is in Vote Leave's campaign a conjuring of ideas of somewhere better; however, not out in the wider world of international tourism destinations, but rather within the enclosure of the UK and the bringing into control those who reside within its walls.

If we recall Buzard's (1993) observations about UK travellers 'exploding' onto the continent at the end of the Napoleonic Wars, along with the feeling of the need to be elsewhere other than the UK in order to feel and practice freedom that was articulated by tourists and expats in Magaluf and Palmanova, the curtailment of this liberty seems to be at odds with a road to freedom. I cannot help but recall the story of a small island to which a usurped king and his young daughter fled. Upon arrival, they find a spirit who has been cruelly imprisoned by an evil witch. The newly arrived Prospero sets the spirit – Ariel – free only to imprison him again for his own bidding.

Acknowledgements

I am grateful to Marc and Ella Roberts for their ideas on the meaning of freedom in the contemporary world that have helped me to clarify some of my thinking for this chapter. I also remain grateful to the tourists and expats who were willing to share their thoughts with me during my various sojourns in Mallorca.

Notes

(1) Oliver Goldsmith. In Arthur Friedman (ed.) *Collected Works of Oliver Goldsmith*, 5 vols, 1966, vol II p. 210 in Franklin 1989: 155.
(2) http://www.voteleavetakecontrol.org/why_vote_leave.html (accessed May 2019)
(3) https://mediatel.co.uk/newsline/2006/10/30/the-sun-cuts-cover-price-to-keep-circulation-up/ (accessed August 2019) This link is provided for purposes of illustrating the image.
(4) https://www.theguardian.com/politics/2016/nov/04/enemies-of-the-people-british-newspapers-react-judges-brexit-ruling (accessed August 2019).
(5) https://en.wikipedia.org/wiki/R_(Miller)_v_Secretary_of_State_for_Exiting_the_European_Union (accessed August 2019).
(6) http://www.voteleavetakecontrol.org/Why vote leave.html (accessed July 2019).
(7) https://www.theguardian.com/politics/2019/jul/04/ann-widdecombe-likens-brexit-to-emancipation-of-slaves (accessed July 2019). It is interesting to note that the language of bound labour was used in opposition to the ideals of Vote Leave when in reaction

to the Conservative proroguing of Parliament the shadow Chancellor John McDonnell accused Prime Minister Johnson of 'treating us [the populace] like serfs' https://www.theguardian.com/politics/live/2019/aug/29/backlash-after-boris-johnson-prorogues-parliament-ahead-of-brexit-live-news (accessed August 2019).
(8) This refers to the idea that one can indulge in as many fleeting, casual sexual relationships as desired.
(9) Peter Stringfellow was a nightclub owner among which was the London-based 'gentleman's club' Stringfellows.

References

Andrews, H. (2009) 'Tits out for the boys and no back chat': Gendered space on holiday. *Journal of Space and Culture* 12 (2), 166–182.
Andrews, H. (2011) *The British on Holiday: Charter Tourism, Identity and Consumption*. Bristol: Channel View Publications.
Andrews, H. (2014) The enchantment of violence: Tales from the Balearics. In H. Andrews (ed.) *Tourism & Violence* (pp. 49–67). London: Routledge.
Andrews, H. (2017a) Mass tourism in Mallorca: Examples from Calvià. In D. Harrison and R. Sharpley (eds) *Mass Tourism in a Small World* (pp. 181–190). Oxford: CABI.
Andrews, H. (2017b) Touring the regions: (Dis)uniting the kingdom on holiday. *Journeys* 18 (1), 79–106.
Berlin, I. (1990) *The Crooked Timber of Humanity: Chapters in the History of Ideas*. London: Murray.
Buzard, J. (1993) *The Beaten Track. European Tourism. Literature, and the Ways to 'Culture' 1800–1918*. Oxford: Clarendon Press.
Cesarani, D. (1996) The changing character of citizenship and nationality in Britain. In D. Cesarani and M. Fulbrook (eds) *Citizenship, Nationality and Migration in Europe* (pp. 57–73). London: Routledge.
Cinnirella, M. (2000) Britain: A history of four nations. In L. Hagendoorn, G. Cseppeli, H. Dekker and R. Farnen (eds) *European Nations and Nationalism: Theoretical and Historical Perspectives* (pp. 37–65). Aldershot: Ashgate.
Franklin, J. (1989) The liberty of the park. In R. Samuel (ed.) *Patriotism. The Making and Unmaking of British National Identity Vol. III: National Fictions* (pp. 141–159). London: Routledge.
Hanefors, M. (2001) Paradise regained Swedish charter tourists visiting the non-ordinary. Unpublished PhD thesis, University of North London.
Jackson, M. (2013) *Lifeworlds. Essays in Existential Anthropology*. Chicago, IL: The University of Chicago Press.
Kelty, C.M. (2017) Liberty and lock-on. The trouble with freedom in anthropology. In M. Lino e Silva and H. Wardle (eds) *Freedom in Practice. Governance, Autonomy and Liberty in the Everyday* (pp. 164–186). London: Routledge.
Lino e Silva, M. and Wardle, H. (2017) Introduction. Testing freedom. In M. Lino e Silva and H. Wardle (eds) *Freedom in Practice. Governance, Autonomy and Liberty in the Everyday* (pp.1–33). London: Routledge.
Lunn, K. (1996) 'Reconsidering Britishness: The construction and significance of national identity in twentieth-century Britain. In B. Jenkins and S.A. Sofos (eds) *Nation and Identity in Contemporary Europe* (pp. 83–100). London: Routledge.
Malinowski, B. (1947) *Freedom and Civilization*. London: George Allen & Unwin.
Mitsui, H. (2017) Becoming 'no one': Muneyoshi Yanagi's theory of freedom in the figure of the unfree craftsman. In M. Lino e Silva and H. Wardle (eds) *Freedom in Practice. Governance, Autonomy and Liberty in the Everyday* (pp. 54–62). London: Routledge.
O'Reilly, K. (2000) *The British on the Costa Del Sol: Transnational Identities and Local Communities*. London: Routledge.

Pagden, A. (1993) *European Encounters with the New World*. London: Yale University Press.
Rapport, N. (2017) The inscrutability of freedom and the liberty of a life-project. In M. Lino e Silva and H. Wardle (eds) *Freedom in Practice. Governance, Autonomy and Liberty in the Everyday* (pp. 34–53). London: Routledge.
Triandafyllidou, A. (2002) *Negotiating Nationhood in a Changing Europe: Views from the Press*. Ceredigion: Edwin Mellen Press.

9 Divisions and Schisms in the Party Space

Anne Storch and Nico Nassenstein

Introduction: Brexit for Fun?

One way of investigating the instability of orders is to create images of misplacement. In a cartoon by the German caricaturists Greser and Lenz[1] (Figure 9.1), everything seems out of place: the motif of the cartoon itself, the drawing, eliciting racist images, for which it was immediately criticised,[2] the language that mimics colonial stereotypes as well as contemporary bureaucratic jargon, and the reference to Brexit in the caption. Brexit, this cartoon seems to convey, is pure misplacement, an anachronistic event that marks the return of all that was hidden underneath. The tropes of colonialism and nationalism return to the surface, as well as the image of a state prior to order – after Brexit, the world is reorganised, but how, and by whom? This is no scramble for Africa; in the cartoon, two African-looking men write a letter in front of a small hut that resembles colonialist European images of African houses. Severed heads are hung under the roof. A hunter pulling an antelope passes by. This is the Global South, prior to colonisation, a stereotype that seemed to have been on the verge of being forgotten for a short time – but here it is again. However, it is a potentially powerful South, as the men prepare an application for inclusion in the European Union (EU): 'Great Master Juncker, herewith we apply for the place in the EU that was recently vacated'. The order of the past, but inverted and disturbing. 'Brexit as an opportunity', the caption of the cartoon reads.

The return of the unspeakable, of history still unmourned, the resonances of a violent order that continues to shape many people's lives around the world are a subtle subtext of this cartoon. Brexit has elicited a particular kind of laughter, bitter and cynical (Koller *et al.*, 2019; Nassenstein & Storch, 2020). The jokes and cartoons that have circulated during the period of the preparations for Brexit often highlight notions of futility, anarchy, anachronism, essentialism and exclusion; these are countered with the production of disturbing, abject images that appear to convey one thing more than anything else – misplaced time and order; a messy present. The inherent disorder of the recurrent images,

Figure 9.1 Brexit as opportunity (Greser & Lenz, 2018)

and the sarcasm found in the cartoon, reflect what the entire Brexit discourse insinuates: a critical turn towards the artificial construction of a unified Europe, EU or Schengen space as a bastion that shields its borders against intruders, refugees and unwanted migrants from the South. What used to be stable is no longer; and those European powers who once drew arbitrary boundaries in the era of imperialism and colonialism across other parts of the world, now face despair over their own policies of inclusion and exclusion. However, Brexit and its effects have not only become evident in European political discourse and in critical cartoons but also in unexpected contexts, such as the tourism strongholds of the Balearic Islands.

The sociolinguistic sites which we explore here, namely two locations of party tourism on the Mediterranean island of Mallorca, S'Arenal and Magaluf, are sites that promise endless fun and possess their own forms of coloniality. While S'Arenal has been a popular destination for party tourists from Germany over at least the past three decades, based on a booming package tourism industry, Magaluf has emerged as a hub for British party tourists, especially from the 1990s onwards. Both places share characteristic features in terms of touristified architecture and excessive consumerism: an agglomeration of multistorey hotel buildings, themed neon light clubs and bars; large numbers of tiny shops selling souvenirs, party costumes and alcoholic beverages; and pedestrian precincts lined with restaurants and shops, which connect the beach and the *paseos* with the club and hotel zones nearby. Andrews (2006) characterises these spaces as almost 'non-places' that serve only for touristic consumption, where one has little option but to look at what is offered and then buy. The consumption of carnivalesque spaces in such party settings is mostly characterised by liminal performances, by 'filling' and 'appropriating' space through nationalist discourse (and a division into a German zone vs. a British zone, and within these zones recurrent and more fine-grained boundaries again), through food- and drinkscapes and through bodies, their sexualisation and, at the same time, their stigmatisation. Visitors' complex knowledge of these semiotic landscapes and their indexical signs intertwine 'contours of a nation' and 'the understanding of a national identity' (Andrews, 2010: 27), made public through flags and dominant soundscapes,[3] with sexualised banality and liminal touristic performances (for instance based on the declared mission of binge-drinking, unrestrained sex, the targeted experience of drug abuse and so forth).

The banality of mass tourism is disguised by the humour and irony used in advertising them; S'Arenal is much better known as *Ballermann* or as Germany's 'seventeenth state' (in addition to its existing 16 states), as a German petty bourgeois colony in the Mediterranean, while Magaluf is *Maga*, infamous for its working-class and binge-drinking image, the name at times also corrupted to *Shagaluf*, due to its sleazy reputation of ubiquitous and tabooless sexual intercourse (Andrews, 2017: 85). However, with new laws enforced on public nudity, sex, alcohol and substance abuse in Mallorca, the party crowd is moving on. Many of those who remain at the island's party sites face the rearrangement of the tourist-based economy and politics and are left with few opportunities available to them – the West African migrants who work as street vendors, lavatory attendants and security personnel, employed on short-term contracts or as seasonal workers. What does Brexit mean to them, to those who have intimately interacted with British tourists and worked for employers who catered especially for British customers? What will be the consequences of Brexit for them, and what are the narratives about British tourists that are circulating at the time Brexit is (maybe) about to become a reality? How will demarcated tourist spaces bound to collective ideas of nations (S'Arenal being proclaimed as German vs. Magaluf as a British territory) be perceived and described by these African actors after a potential Brexit?

This chapter is based on fieldwork conducted around the party hotspots of S'Arenal, and to a lesser extent also Magaluf, between 2016 and 2019 in the context of an interdisciplinary research project on tourism, language and migration in Mallorca. We focused predominantly on analyses of fleeting relationships and on the collection of 'small stories' (Storch, 2017), collected in washrooms, along the beach promenade, in shops and in restaurants. Small stories, as a category of personal, at times incoherent observations and evaluations of quick encounters between tourists, tourism workers and hosts, also play a role in other current approaches to migration and migrants' narratives in sociolinguistics. Another common method during our fieldwork periods was extended walks (Horvath & Szakolczai, 2018) with Senegalese beach vendors (cf. Nassenstein, 2017), both during their working hours and in their free time. While fellow sociolinguists often tend to reject more personal, less data-oriented and unconventional methods of collection, our work with largely undocumented migrants in precarious conditions made these very non-mainstream approaches a prerequisite in order to keep interactions at a safe level and guarantee our interlocutors' safety.

Segregation at the Beach: On National(ised) Spaces and Marginalisation in Between

The two sites, S'Arenal and Magaluf, also have in common an undebatable hybridity, both in terms of class and identity performances and in terms of the geopolitical origins of those who stay, work, consume and play there. Tourists and migrants alike are diverse in terms of their biographies, socioeconomic backgrounds, origins and so forth; stereotypical binary constructions of the (British or German) *tourist* and the (West African) *migrant* are rather reflections of unequal encounters in the touristic setting, which itself is based on socioeconomic inequalities. However, the modalities and motivations of their multilingual repertoires and of their language use *do* diverge: Language is important to move forward for some, but not for all who populate these sites. Whether one depends on broadening and adapting one's communicative repertoire or not is one of the decisive differences between privileged and marginalised players in this setting. But communicative practices, literacies and language ideologies are not the only parameters which determine prospects and opportunities; the materiality of the existences in the party space are more visible ones, and they illustrate clearly which forms of segregation actually shape these sites and thus provide the background for a Brexit discourse that differs only slightly from that in the above-mentioned cartoon.

The division into one German and one British party area is not as clear and homogeneous as it seems at first sight: more fine-grained lines of demarcation are reproduced within these spaces. In Magaluf, one sideroad is populated by partying tourists from Sweden, Norway, Denmark and Finland, with an abundance of Scandinavian flags at the entrances to bars; in S'Arenal, short stretches of the *playa* are British and Dutch dominated and are rarely frequented by German tourists. This kind of 'fractal recursivity' (Irvine & Gal, 2000), a reproduced division (of geographical origin) on another level (here indexing different nationalist identities and different ways of doing vacation) does not reduce the 'contours of a nation' (in Andrews' terms) but actually strengthens them through the emergent discourse on striking differences and similarities: in some online forums, one finds stories of young British tourists complaining about 'horny, drunk-as-fuck Scandinavians', in others complaints from Swedish families who are disgusted by the noise and public sexual behaviour of British youths; again others narrate incidents of shocked British families right at 'the moment when you realise you're in a German hotel!' (as found in various travellers' reviews online). A group of elderly Germans in their 60s and 70s, who shared the history of their travels in Mallorca with us, based on almost annual visits, barely noticed that their voices changed when speaking of the four to five bars at S'Arenal 'where the British are' and where one could still sip from a 'way too sweet Sangria in the Dutch bars' (extracted from interviews in 2017).

The internal ghettoisation and reproduced differentiation (Swedish vs. British, Dutch vs. German and so forth) seems to be a major driving force for the success that charter tourism resorts like S'Arenal generate: the Other is a permanent part of the performance, and cultural differences increase Britishness, Germanness or a pan-Scandinavian identity. Andrews (2017: 90) analyses the construction of British tourist identity in Magaluf (see Figure 9.2) and concludes that 'while there is unity in the face of the foreign other, within that unity there are more localized ideas of identity'; these further fracture the layers of tourists' identity constructions and contribute to a picture where various opposing semiotic signs stand at the centre of the public discourse.

But how do Senegalese street vendors, Nigerian lavatory attendants and other apparently invisible and seemingly liminal actors view these politics of identity and boundaries, particularly during the rise of an imminent Brexit and its after-effects?[4] And which changes, both socio-economic and political, will a more 'dis-United Kingdom' (Andrews, 2017: 90) yield in one of its citizens' most-frequented tourist locations? It becomes evident that the typologies of categorisation reach way further than only from one party street to another, separated by different-looking European flags. In the margins around European tourist identifications and indexical signs of Britishness, Germanness, etc., there are Asians, Africans and Latin Americans, their languages and diverse patterns of communication. At the beach, one of the most obvious aspects of the party tourism industry are the many shops that offer motto T-shirts, motto beach towels, souvenirs, beach clothing and often party costumes as well. Many of the owners and employees are from India but have lived for decades in Mallorca, as well as in other parts of Spain and Europe. The shops are located near the clubs and bars, and many of the owners have a good knowledge of the languages spoken by tourists, especially German and English. Some of the motto T-shirt traders are able to speak German to such an extent that they design their own collections of T-shirts for their German customers. Communication among shop owners and employees is often in Hindi, while contact with

Figure 9.2 Performed Britishness (Photo from Storch, 2018)

other inhabitants of the area tends to be maintained in Spanish. Catalan/Mallorquin does not play a salient role, many among them say, since it is only spoken by those who originate from the island and who are claimed not to be interested in having any relationship with migrants. As well as these shops, there are spots in front of the clubs where street vendors, mostly West Africans, sell not only sunglasses, watches and bracelets, but also illegal substances of various kinds. In S'Arenal, the street vendors are busy throughout the day, browsing along the *paseo* and beaches during the day and standing in front of the clubs at night. In Magaluf, many only begin their shift at night, when more customers are around. Some of the men have spent years around the beach areas, while others have been around for shorter periods of time. Communicative repertoires match each other, as most of the street vendors speak German or English and Spanish, among other languages such as French, Wolof, Mandinka and Fula (all languages spoken in Senegambia and beyond). Yet, there is little interaction between players who see themselves as belonging to different social groups, against a background of the dominant indices of German- and British-targeted touristic markets.

Chinese supermarket entrepreneurs form another distinct group, supplying a large number of diverse customers not only with low-priced items to sell on the streets, but also with items used in other everyday contexts, such as household tools and cheap clothing. Communication usually takes place in Cantonese among staff, and in Spanish in interactions between staff and customers. At the beach itself, there are Chinese masseuses who offer their services, working in small groups and speaking to the tourists in whatever language might be needed, based on a patchwork repertoire of numerous languages.

Other players at the beach are sand castle artists, who are mostly from Romania; vendors of cold drinks, who come from Andalusia; a group of capoeira artists who originate from Brazil and who perform at night on the *paseo*; and a large group of women, mostly from Nigeria and usually seen as sex workers, although many work as household helps, shopkeepers and hotel staff. Then there are food vendors and butchers from Morocco and businesspeople from Argentina, operating cafeterias, small shops or whatever might come along.

Language is plentiful here, and diverse, going beyond the dominant voices that resonate in German and English. Those who have little choice in deciding what they want from being in this place need to be more versatile than others, bridging the gap and making contact with tourists and future employers, and with local authorities when necessary. Segregation works along various lines – one's origin and group affiliation, which often follows colonial stereotypes ('Indian', 'African'), as well as the material context of one's work (e.g. owning a shop, being employed with a contract, working in the street or without shelter, as an undocumented migrant). Segregation at the beach is visible and brutal at both party sites and yet it is an undisputed reality of them. Tourists make use of it in terms of establishing routines, such as buying a cold drink from a Romani person cheaply at the beach, looking out for affordable blow jobs from Nigerian women at the *paseo* by night, playing games with or abusing a street vendor who might be without legal paperwork and therefore unable to complain. Segregation, not only between Germans and British, or between Scottish and English, is part of the party and is profitable for some, particularly for many of the tourists.

Over the three-year period during which we explored language practices, mostly in S'Arenal, but also in Magaluf, we were particularly interested in language acquisition and language ideology among the West Africans in the party zone. Most of the beach vendors we interacted with explained that they would learn language(s) from their interactions with tourists and others, at the beach and in the streets. Spanish was learned both in the streets and through classes offered by welfare organisations or the state. More recently, migrants have developed an increased interest in taking part in English and German lessons at language schools in the area, as the new tourism policy on the island requires more highly qualified staff.

But the main resource remains the street, where those who wait for customers need to attract attention. Based on our conversations, an interesting estimation was that language was usually considered hard to learn when the social environment was uninviting or hostile, and, in contrast, easy to acquire, whenever those speaking it were interacting more freely in less hostile surroundings. Since German party tourists regularly included the predominantly Senegalese street vendors in their party performances in the streets late at night, it was relatively easy for them to pick up words, we were told. This was different with British tourists in Magaluf. To a great extent, they were not interested in any playful performance, as various Senegalese interlocutors told us, and would thus keep their conversations with the street vendors to a minimum. English was 'a bit hard to learn', they claimed, because it was more socially inaccessible than the other languages that surrounded the migrants. The same applied to the Dutch language in S'Arenal, as some Senegalese conversational partners emphasised: most Dutch people spoke excellent English or German and the vendors would thus lack the opportunity to learn Dutch; moreover, Dutch had 'strange sounds', and our interlocutors would demonstrate this impressively by mimicking a hissing cat.

Inefficiency in communication was translated into inefficiencies in integration: the Germans integrated themselves well in some places, as Senegalese migrants said, but needed more moral and cultural education, as their transgressive behaviour at the *Ballermann* remained problematic and offensive to many of the mobile tourist workers. Integration as the social ability to adapt to different sociolinguistic contexts and demands was also a key motif in Brexit discourse (as based on our preliminary research on this matter). The relative lack of interest among British tourists in 'integrating', as well as the pre-existing patterns of segregation at the beach and party space, were seen as a 'Brexit normality' that predated the referendum. 'What will change after Brexit?' was a question that elicited shrugged shoulders: 'Everything is already separated' (meaning 'divided' and most probably translated literally from French). There was a hint of disappointment in this statement.

Language has its particular indexicalities within the party space, because it is the most important tool and medium not only for work, but also for migration. At the beach, in *Maga/Shagaluf* and the *Ballermann*, the key competence for success as a street vendor or temporary worker in the tourist industry is one's ability to adapt and enlarge one's linguistic repertoire. Being able to speak some of the tourists' languages, in addition to a solid basis in Spanish, can be measured as equivalent to establishing relationships by taking part in language games, carnivalesque performances and play, all of which result in relatively reliable business during the peak tourist seasons. Being dressed in party hats and soccer shirts, wearing plastic sunglasses and dozens of bracelets alone does not attract enough attention, as most visitors to these party zones dress up in carnivalesque ways, with motto T-shirts and colourful costumes. It is, rather, one's proficiency and versatility in English and German, and to a lesser extent also in other European languages, that helps a vendor to get close enough to customers. Ahmed, a Senegalese migrant who has lived and worked in S'Arenal for about five years, explained: 'This is a purely German economic zone. Learning is easy, but not of good quality, I would not be able to [do so]. You will speak, you will work, you will know some, you will work…' (Ahmed, interview 2017).

Another Senegalese man addressed the topic of language games in the party hotspots – such as receiving nicknames, repeatedly using slurs and sayings learned from the tourists, as well as from listening to the party songs that are audible everywhere and played on an infinite loop (see the next section): '*Ah Helmut ça va? C'est nécessaire de parler beaucoup de langues. Helmut, c'est la farce, c'est un jeu, c'est rien de grave. Hellmut, Dunkelmut … hätte hätte Fahrradkette, heute billig morgen teuer, hundert Jahre Garantie …*'.[5]

Seasonal or short-term employees, who work as security personnel, waiters or lavatory attendants in the large clubs and discotheques, are requested to speak at least enough English and German to assist disoriented drunk customers. Yet, the relative reluctance of tourists to engage with West African migrants or to spend more time with them than absolutely necessary, helping with language or showing gestures of generosity (for example, by buying more souvenirs than initially needed, or tipping a vendor), was considered by many Africans working in the party tourism zone as a sign of a more general withdrawal: on the eve of Brexit, life on Mallorca gets continuously more difficult for them, while Europe is closing its gates. Brexit and the renegotiation of borders may also bring back memories of colonial times and of arbitrary lines of demarcation that caused socioeconomic hardship, aggravated dependencies and stole individuals' hopes for better lives. The Mallorcan case, the imminent Brexit and a drastically changing flow of tourists with fewer customers, more competition, stricter regulations and laws may therefore also contribute to a general change of perspective, as will be explained further below.

One sign of this change is the inhospitable signpost placed at the entrance of the widely known party bar Bamboleo (in S'Arenal), informing clients in German and French (not Spanish or English) that the toilets are reserved for customers, indicating that (Francophone) Africans are not permitted inside (Figure 9.3). While translation (into the language of a group as an act of facilitating communication) is usually a hospitable action, allowing for quick comprehension, here translation into a language almost uniquely spoken by marginalised and excluded actors turns into a stigma.[6] The motivation for speaking less French and gaining a more solid basis in German and English (for business) as well as Spanish (for matters of integration) among African migrants is therefore self-explanatory.

Another sign of the changing practices of exclusion and inclusion seems to be the recently positioned massive concrete road blocks that were placed in the streets in the winter of 2018, and which resemble those

Figure 9.3 Visitors only (S'Arenal; photo Storch, 2017)

in various European metropoles. At S'Arenal, they can be understood as recent landmarks that both generate and reduce fear at the same time. The Senegalese street vendors we interacted with, most of whom are Muslims, expressed their lack of understanding for such a seemingly hostile gesture, stating: 'This is a quiet place. No terrorists here... No idea why they put these things. Because of Barcelona I think' (Interview, February 2018).

The apparent hostility that dominates the professional encounters of tourists with many street vendors and other marginalised players at the beach is an issue that has frequently been addressed. A 'lack of proper education' was mentioned as a possible explanation; drugs and alcohol, as well as continued excess during vacation, as another. Disappointment with the lack of access to English (more than German), however, appears to have more complex reasons than this. In her work on *Senegal*

Abroad, Maya Angela Smith (2019) emphasises the relevance of language ideology:

> English has a much more prominent position than Italian [or Spanish] in the Senegalese cultural imaginary. At the same time, the attitudes appeared less complex than those concerning French. The prestige of English relative to French might seem surprising given its colonial implications – English was also a colonial language, evidenced most strongly by the fact that the Gambia, which bisects Senegal, is an English-speaking country and former British colony. However, English in the Senegalese imaginary is not viewed through the same postcolonial lens as French most likely because it was not the colonizing language of Senegal. I received barely any negative opinions about the English language, which provoked little disdain from a postcolonial perspective and garnered respect because of its global influence. (Smith, 2019: 48)

Yet, both Francophone and Anglophone West African tourism workers in Magaluf and S'Arenal are confronted with the continuing marginalisation of African Englishes and literacies, which is part of what happens in the black box of migration (Lindtner, 2018: 149–155). The segregation of businesses and presences along the lines of colonial geopolitics at first sight seems not to affect language, which is strikingly diverse here. But there is a conspicuous absence of Catalan/Mallorquin from most migrant repertoires. 'The locals don't want to have anything to do with Africans' was a frequent view expressed by West African migrants concerning their situation away from the tourist spaces. A Nigerian woman who worked as a cleaner in a club in S'Arenal stated that, at first, her husband and many other Nigerian men had to spend their nights in the graveyard, because nobody would rent out apartment space to Black people. The owner of a prosperous taxi company, Festus Badaseraye, who migrated from Nigeria more than two decades ago and is now a holder of Spanish citizenship, has filed a lawsuit against policemen in Palma for the racist treatment he had to endure when attempting to register a new car.[7]

As a result of such experiences and of recurrent discourses on the linguistic difficulties African migrants face in Mallorca (and elsewhere),[8] many speakers of English – Nigerian, Senegalese and other African migrants – expressed their wish to continue their migration from Spain to Britain, where African diaspora communities were considered to be better established, and represented in much larger numbers and over a longer period of time. Economic success in the party hotspots was considered one possibility for establishing a more economically secure life elsewhere in Europe. Rather unexpectedly, in summer 2018, against the background of many individuals' firm wish to leave Spain, Brexit emerged as a central reason for the economic failures of African immigrants, with the party area changing into a more competitive and exclusive space. Additionally, the increasing gentrification and upscaling of the old buildings and establishments incipiently reduced the options for African migrants in these areas. New obstacles and borders emerged next to old ones – borders that appear as little permeable as the old ones, despite their unconventional architecture.

Sound as a Border

The party beach has an ambiguous character, as it is both a site of freedom and transgression, and a site of regulation and consumerism, where transgression is scripted, the violation of taboos sanctioned and access is never free but has to be paid for. While tourists oscillate between moments of transgression and heavy fines when violating regulations of good conduct, the African actors are confronted with multilayered boundaries. In addition to the anti-terror concrete blocks and the restrictions on using specific restrooms (indicated in French), there is a clear policy on where an African street vendor is allowed to engage in business transactions and conversations with tourists, and where not. In the central party zones, i.e. the relabelled streets *Schinkenstraße* and *Bierstraße* of S'Arenal, they are not allowed to cross an invisible line, the separating line between street and sidewalk (where tourists sit, chat and drink beer). Even though Africans are integrated in performances of party tourism, they are not permitted into the actual beer parlours and clubs. The Senegalese street vendors and Nigerian sex workers are supposed to stay outside, where they remain vulnerable to practices of social ostracism and violence.

However, apart from visual and architectural boundaries that serve as markers of exclusion, the sonic performances of ostracism can also be perceived as borders (cf. the example of labelling and naming migrants *Helmut*, above) that draw clear boundaries between groups of different status and turn social inequalities into aural experiences. Besides language games and mocking strategies (imitating 'Broken German' or defective and limited English, producing a ridiculed copy of an African language, which also takes place in social media etc.[9]), these boundaries consist of blaring tourists' voices and, more substantially, of loudspeakers that play the above-mentioned party hits, many of which reproduce stereotypes about street vendors, female restroom cleaners, sex workers or drug runners, and relate them to the migrants' pigmentation, their identity as foreigners from the Global South or their marginal(ised) status – further marginalising them through these songs. In the party-dominated spaces, sound thus functions as a clearly demarcated border, and turns language into a burden – hard to deal with and almost impossible to avoid. The songs reveal very directly that Africans are targeted, operating on similar racist tendencies to those in the Greser and Lenz cartoon, with clear messages of ostracism and rejection. One example is Honk's (2017) song

'*Hallo Helmut (andere Farbe)*', in which blackfacing, instrumentalised racism and party tunes work together to create a boasting sonic medium of exclusion that is suitable for the large party arenas.[10] The message – and the communicated border – is presented point-blank, just as it is in the subsequent (2018) song '*Klauhurensong*', which addresses Nigerian sex workers, referred to with the derogatory label of *Klauhuren* 'whores who are thieves/stealing hoes'.[11] In these cases, the sonic borders are violently presented and their steady repetition in the party spaces marks African migrants as outlaws.

Further, the sonic boundaries booming from loudspeakers serve not only to exclude marginalised social groups such as West African migrants, but also to construct spaces of apparent refuge and privacy, and of quick pleasure and long-sought transgression, for package tourists. Sound serves as an invitation for some (to party, to go on a pub crawl, to find distraction), but not for others, who are categorically excluded. The frequency with which new hit songs are played throughout the party area of S'Arenal (while in the British-oriented equivalent Magaluf a completely different style of music is prevalent) makes retreat impossible for the listeners affected, whose linguistic knowledge is often – especially with regard to the frequency of the played songs and their visualisation as print extracts on motto T-shirts – sufficient to capture the sense of what is being sung.

The multiplicity of borders in such highly diverse settings becomes evident in the observation that borders are not only articulated through laws and regulations, written signposts and rejected paperwork or lack of administrative support, but also through sound. Street vendors who share costly apartments with several other colleagues with similarly precarious backgrounds, and who are confronted with sonic encroachments throughout their daily work, cannot claim any private space, nor can they construct a confined area where a personal and emotional life can take place. Street vendors such as our interlocutors are therefore forced to negotiate sound permanently as a salient border.

Where the pressure to perform in languages that are learned out of economic necessity and at times based on painful modes of acquisition is not present for a moment, and where no party music can be heard, a sense of privacy may be felt (e.g. along the silent cliffs of the coastline a few kilometres behind the busy beaches, in Senegalese coffee shops or in the run-down neighbourhood of Son Gotleu, where many Africans live), while noisiness seems to be a marker of the workplace – the street and beach areas, and in front of the beer parlours. Acoustic borders as decisive – yet invisible – indication lines of inclusion and exclusion are therefore strict boundaries (just as much as the linguistic line of division between S'Arenal and Magaluf) and locate the African migrants somewhere in between the dominant European languages and the British and German party neighbourhoods, beyond discourses of nationalist belonging – or rather as belonging neither to continental Europe (and its S'Arenal) nor to a separate United Kingdom (and its Magaluf).

Conclusion

Not only in its constant occurrence in political discourse but also in economic analyses of all kinds – related both to the tourism industry and to many others – the Brexit discourse has been transformed into a variety of material forms. It has primarily resulted in the emergence of Brexit-specific souvenirs in some of Magaluf's gift shops. Tourist settings offer a wide range of material holiday memories. Specific artefacts appear in an abundance of variations and modified shapes, such as sexualised souvenirs, whose emblematic and commodified value (see Traber, 2017) is intended to trigger hilarity, embarrassment, curiosity and consequently a quick purchase. Very often, as already indicated, the shops in Mallorca are run by Indian or Pakistani T-shirt sellers and printers, by Chinese companies or extended families, and only occasionally by Spanish business owners. In Magaluf, some of the shop clerks have come up with a *Brexit roll*, toilet paper on which Euro banknotes are printed (Figure 9.4).

This creative item conveys the message to customers that the Euro, after a successful Brexit, may soon only be worth using to wipe one's behind, at least from the perspective of a British post-Brexit witness. But a different reading is also possible: a successful Brexit could result

Figure 9.4 Brexit toilet paper purchased in Magaluf (Archive Storch)

in a shortage of basic items needed in daily life, such as toilet paper (it may thus be safer to buy ahead); here, Brexit and the roll of toilet paper would represent fear and lack of opportunities. We also observed other forms of embodiment and masquerade in Magaluf, including tourists' costumes and also, to some extent, the ostracising practices in Honk's video (see above). After purchasing (stereo)typical items that are associated with 'Germanness', British tourists impersonated German tourists: the exoticism and orientalism of continental Europeans (Germans) was here paired with the Southern stereotypes they perform at the party space, as the costumes were combined with blackfacing practices. These carnivalesque jokes can also be subject to very different readings; this ambiguity seems to range among the few stable and constant attributes that the current Brexit discourse has so far brought along.

The question 'Brexit as opportunity or change of perspective?' therefore yields diverse and debatable replies. Brexit may be perceived not only as an opportunity but could equally lead to a change of perspective in African migrants' views of Europe, or Britain, and in regard to the ubiquitous divisions in touristified spaces, between the British and other tourists. Brexit might, then, seen from the perspective of African migrants to the Balearic Islands, repeat the colonial 'cage-rearing' that hindered generations of colonised and exploited individuals from moving freely, or claiming mobility. Moreover, Brexit already has and undoubtedly will still contribute to the changing economic conditions of many West African migrants in different parts of Europe. The Nigerian cleaning woman who expressed her wish to migrate to Britain (Interview, 2018) has since decided to stay in Mallorca. Brexit and its implications are complicated, and the news is dominated by pessimistic voices; potentially aggravating one's own already precarious situation is thus not an option for many. Some of the extreme political positions and tourists' ostracising practices of drawing boundaries between themselves and the African migrants may become more solid due to Brexit. The schism between tourists and migrants may, again due to the economic changes in the tourism business, thus turn into a still more violent line of division. The cartoon discussed at the beginning of this chapter also allows for a racist distortion of a changing Mallorca that will join the EU as an African colony.

Notes

(1) https://www.greser-lenz.de/witze-archiv?album=8&gallery=52 (accessed 1 March 2019).
(2) https://uebermedien.de/32301/verletzende-stereotype-rassistische-karikaturen-von-greser-lenz/ (accessed 1 March 2019).
(3) Ryan (1995: 211) narrates an incident where these dominant national(ist) incorporations of British identity, when disturbed by 'entertainment in the hotel [...] oriented towards longer stay Spanish pensioners', were perceived as a 'rude surprise' and triggered both anger and violent reactions.

(4) However, it must be mentioned that questions around one's nationality play a role not only in European tourists' discourse, but also in African migrants' discourse on their country of origin. One's true identity and nationality is often concealed; Malians, Guineans and Gambians all claim (for reasons of simplification but also as a strategy of anonymisation/protection) that they originate from Senegal. One African street vendor claimed that he originated from Somalia, while others indicated that he was from West Africa (and he communicated in Wolof).
(5) Translates as 'Oh Helmut (nickname for West African street vendors), how are you? It is necessary to speak many languages. Helmut is just a joke, a game, nothing serious. Light-*mut*, Dark-*mut*... If, if only if, cheap today, costly tomorrow, warranty one hundred years'.
(6) Also, the variety and specific idiolectal realisation of French commonly indicates a difference in status among immigrants. Senegalese street vendors across Europe are often ostracised because their French is judged to be 'defective' and 'not native', while proficient French would often be equated with White race. A White French speaker would then be labelled *Français de souche* (Smith, 2019), indicating that being French and being Black are still mutually exclusive, a long time after Fanon first wrote about this in 1952.
(7) https://www.ultimahora.es/noticias/local/2017/03/24/256756/govern-abre-diligencias-tras-denuncia-por-delito-odio-linguistico.html (accessed 20 February 2019).
(8) http://manteros.org (accessed 26 February 2019).
(9) For these offensive practices, see also the Instagram account #helmutbestermann, the official Instagram platform of the widely known bar *Bierkönig* (accessed 3 March 2019).
(10) https://www.youtube.com/watch?v=TKfxKktQGKY (accessed 20 March 2019).
(11) https://www.youtube.com/watch?v=P2W7G7hYVKg (accessed 17 March 2019).

References

Andrews, H. (2006) Consuming pleasures: Package tourists in Mallorca. In K. Meethan, A. Anderson and S. Miles (eds) *Tourism Consumption and Representation* (pp. 217–235). Wallingford/Cambridge, MA: CABI.

Andrews, H. (2010) Contours of a nation: Being British in Mallorca. In J. Scott and T. Selwyn (eds) *Thinking Through Tourism* (pp. 27–50). Oxford: Berg.

Andrews, H. (2017) Touring the regions. (Dis)Uniting the kingdom on holiday. *Journeys* 18 (1), 79–106.

Fanon, F. (1967) *Black Skin, White Masks*. London: Pluto.

Horvath, A. and Szakolczai, A. (2018) *Walking into the Void*. London: Routledge.

Irvine, J.T. and Gal, S. (2000) Language ideology and linguistic differentiation. In P.V. Kroskrity (ed.) *Regimes of Language* (pp. 35–83). Santa Fe, NM: School of American Research Press.

Koller, V., Kopf, S. and Miglbauer, M. (eds) (2019) *Discourses of Brexit*. London: Routledge.

Lindtner, E. (2018) *Zwischen Nigeria und Europa. Schicksale von Migration und Remigration*. Vienna: Promedia.

Nassenstein, N. (2017) Une promenade linguistique with a Senegalese street vendor: Reflecting multilingual practice and language ideology in El Arenal. *The Mouth* 2, 80–95.

Nassenstein, N. and Storch, A. (eds.) (forthcoming) *Swearing and Cursing*. Berlin: de Gruyter Mouton.

Ryan, C. (1995) Learning about tourists from conversations: The over-55s in Majorca. *Tourism Management* 16 (3), 207–215.

Smith, M.A. (2019) *Senegal Abroad*. Madison, WI: University of Wisconsin Press.

Storch, A. (2017) Small stories. *The Mouth* 2, 98–117.

Traber, J. (2017) Der Verkauf von Verkehr. *The Mouth* 2, 60–76.

10 Post-Brexit Tourism and the Commonwealth Reimagined

Marcus L. Stephenson and Shaun Goldfinch

Introduction

A post-Brexit Britain era opens up possibilities for strengthened relationships with members of the former British Empire, and with current members of the Commonwealth of Nations. This includes expanding already important trade links and cooperation across a range of initiatives. It also includes tourism, drawing on legacies of language and culture, and shared histories. The Brexit movement itself draws on a politics of nostalgia, particularly a view of a Britain once separated from the rest of Europe and in control of its own destiny – to 'Take Back Control' in the language of pro-Brexit campaign advertising; but also a nostalgia for a time when it was the first-rank power controlling a quarter of the world. Lacking a central role in the politics of the European Union (EU) may give Britain cause to look elsewhere not only to find levers of influence, but also to locate new markets for goods and services and capital potentially excluded from Europe. It also opens up opportunities for new and/or strengthened relationships, and new and revitalised (albeit perhaps imagined) communities around shared values, shared institutions and shared language.

This chapter provides an overview of the Commonwealth, and outlines the so-called Anglosphere which includes a common language, shared values and ethnic affinities. The Anglosphere as a political project can be seen to foster common national identities associated largely with the former 'white dominions', as well as projecting Britain's role in a refocused global order. We examine ways in which tourism and travel can facilitate connexions between the Anglosphere, Britain and the broader Commonwealth, including heritage and royal tourism, and Britain as a destination and tourism brand identity. The chapter then highlights how contemporary forms of heritage can conceal contested histories, shielding tourists from confronting past and unpleasant and uncertain events. The latter part of the chapter, however, unfolds the ambiguities of Commonwealth tourism by showing that although the tourism industry can obscure and obfuscate, in other situations it can be genuinely informative. As such, tourism can open new paths for constructive engagement across the Commonwealth.

The Commonwealth of Nations

The Commonwealth of Nations (henceforth the Commonwealth) was established, or more correctly renamed, in 1949 and currently has 53 members, with a combined population of around 2.2 billion, accounting for 20% of global trade, around 14% of global gross domestic product (GDP) and 20% of the world's land area (The Commonwealth, nd). Not all former British dependencies are members, with notable exceptions being Burma/Myanmar, Ireland and former British protectorates in the Middle East. Nor are all current members former dependencies of the British Empire: Mozambique (a Portuguese colony until 1975) joined in 1993 and Rwanda (Belgian administered until 1962) joined in 2009. The range of economic development and size among members is considerable: from the rich nations of the UK, Canada, Australia, Singapore and New Zealand; middle income but huge economies such as India; and tiny, fragile and aid-dependent low-incomes states such as the South Pacific islands of Tuvalu and Nauru.

Queen Elizabeth II is the head of the Commonwealth and head of state of 16 members. The Commonwealth Secretariat offers a range of services and research in development and institution building to member states. Every two years, the Commonwealth Heads of Government Meeting is held in different locations, bringing together the political leaders of the constituent states, while other fora bring together politicians and public servants on various issues. The Harare Declaration of 1991 committed the Commonwealth to the principles of democracy, good government, human rights, the rule of law and sustainable economic and social development. Members not meeting those objectives can face suspension, as was twice the case for Fiji and Pakistan, and once for Nigeria; though all have since been readmitted. Zimbabwe withdrew from the organisation after its suspension in 2002. One of its most obvious symbols is the four-yearly multi-sport Commonwealth Games, attended by 71 nations and administered territories, and attracting a claimed TV audience of 1.5 billion in 2018 (The Commonwealth, nd).

Britain's military ties unwound outside Europe to some extent after the Second World War, particularly after the fall of Singapore in 1942 and Indian independence in 1947. Yet strong links remain with the Anglophone democracies and with some Commonwealth members, with recent moves to strengthen ties, particularly in the face of Chinese assertiveness in the Asia-Pacific. The Commonwealth remains a key recruiting ground for the UK military, providing around 4500 personnel, with 1400 Fijians making up the largest contingent, and with Commonwealth citizens able to join without a residency period. Canada and the UK are members of the North Atlantic Treaty Organization, while Australia and New Zealand drew closer to the United States and signed the ANZUS pact in 1951. The South East Asian Treaty organisation included the UK, Australia and New Zealand, until its dissolution in 1977. Commonwealth forces defeated communist insurgents in the Malayan Emergency, and Indonesia in the Konfrontasi of the 1960s; and the UK, Canada, Australia and New Zealand are part of the Five Eyes intelligence alliance, along with the United States (O'Neil, 2017). The Five Power Defence Arrangements, established in 1971, are maintained between the UK, Singapore, Malaysia, New Zealand and Australia, which conducts joint armed forces operations and sees small UK contingents maintained in Singapore and Malaysia. Australia and the UK formalised bilateral military cooperation in 2013, and British bases are maintained in Kenya, Brunei, the Falklands, Sierra Leone and Cyprus. Other bilateral ties between Commonwealth members include Australia–India, Australia–Malaysia and Canada–Caribbean agreements. These, along with training provided by British and other Commonwealth forces to developing and other nations, constitute something of a Commonwealth military culture.

Until the 1950s, the Commonwealth was more important to the British economy than Europe, but the UK reoriented itself towards Europe in the post-war era, particularly in the build up to, and immediately after, Britain's membership of the European Economic Community (EEC) in 1973. This should not be overstated: the EU currently accounts for 55% of UK exports and 53% of its imports (Ward, 2019). Australia, New Zealand and Canada, initially strongly integrated into the British market, directed trade elsewhere. The Commonwealth remains a small but significant trading partner for the UK, accounting for 9% of exports and 8% of imports in 2018; with India, Canada, Australia, Singapore and South Africa accounting for 69% of this total (Ward, 2019).

Much has been made of the potential for re-engagement with the Commonwealth. Linking diasporic communities within the Commonwealth is claimed to encourage trade, tourism, remittances, investments flows and knowledge transfer (Business Times, 2018; Grenade, 2016, 2018), with UKIP/Brexit Party leader Nigel Farage arguing that exiting the EU would create opportunities for linking with worldwide partners, 'starting with our friends in the Commonwealth' (Dilley, 2016: np). The Secretary of State of International Trade, Liz Truss, claimed that a free trade agreement could be reached with Australia within months of Brexit (ABC, 2019). Prime Minister Boris Johnson also discussed free movement of labour between the UK and Australasia, claiming that 'we British are more deeply connected with the Australians – culturally and emotionally – than with any other country on earth' (Namusoke, 2016: 469). Moreover, what has been called the narrative of 'Global Britain' connected through free trade and other networks to a world beyond the putatively narrow and sclerotic confines of the European market – and facilitated by new information and communication technologies – was a strand in pro-Brexit campaigning. It sought to make the Brexit project 'more palatable to a wider "liberal/globalist" constituency' and at the same time provide a nostalgic turn to a 19th-century economic free trade order, led by an apparently benign Britain, and overseen by the Royal Navy (Sykes, 2018: 142).

Memory selects, rarefies, misrepresents and simplifies. Nostalgia is a process of forgetting as well as remembering. The Brexit project can treat the sometimes catastrophic unrolling of Empire and the post-war economic crises of Britain as if they never happened. Forgotten too is the perceived need to join with the rest of Europe and the European community with the first UK application in 1961, twice thwarted by Charles de Gaulle's 'non'; and the monumental and perhaps irreversible unwinding of Commonwealth relationships that EEC membership entailed. Beliefs that the Commonwealth can replace lost markets of the EU verges on fantasy. Nonetheless, the complex motivations and obfuscations of Brexit politics do not mean that a re-engagement is entirely without merit. This chapter will, therefore, argue that to simply reject such claims is to caricature what are genuine communities and opportunities, with tourism itself playing a role in engaging with the Commonwealth.

Deciphering the Anglosphere

Beyond – albeit largely including – the concept of the Commonwealth is the so-called Anglosphere (Mycock & Wellings, 2017), Anglo-cluster (Ashkanasy *et al.*, 2002) or Anglo-Celtic world. It is unclear where the boundaries of these lie, though English language is a binding element. Drawing on Hofstede's (2001) cultural dimensions, Ashkanasy *et al.* (2002) see the Anglo-cluster including the United Kingdom, Canada, Australia, New Zealand, the United States and white South Africa, with these countries sharing key cultural traits to the extent of constituting a more-or-less distinct cultural grouping. Bennett (2007: 80) conceptualises the Anglosphere as:

> the United States and the United Kingdom. English-speaking Canada, Australia, New Zealand, Ireland, and English-speaking South Africa are also significant populations. The English-speaking Caribbean, English-speaking Oceania, and the English-speaking educated populations in Africa and India constitute other important nodes.

The Anglosphere's key members are former British possessions, even if the United States and Ireland are not members of the Commonwealth. As a political project, its foundations are in Great Britain's late 19th-century attempt to establish an accord between itself the 'Mother Country' and the (white and more or less self-governing) 'Dominions', which were established by settler colonies and 'people of British stock'; indeed 'Better

Britons' (Goldfinch & Mein-Smith, 2006; Mycock & Wellings, 2017). Its roots, however, originate in the English Empire; perhaps to Norman England's first tenuous colonisation of Ireland in the 12th century, or the later more successful Tudor colonisation.

Moreover, the Anglosphere shares more than just language and culture, with shared institutions of rule of law and judicial independence, common law-based legal systems, constrained states, aspirations towards free media and Westminster-derived political systems. With the United States somewhat of a divergence – albeit with its constitution derived from an interesting reading of the British 18th-century constitution and British philosophers – the United States still uses a common law system and draws on lessons from elsewhere, including a constrained state. This sharing goes beyond institutions adopted at colonisation and includes a history of shared policy, shared judicial findings and shared courts, such as the preeminent role of the Privy Council as the highest court of appeal continuing into the 21st century for New Zealand and some Caribbean nations (Goldfinch & Mein-Smith, 2006). Modern democracy is sometimes seen as an Anglo-American initiative, drawing on 'a tradition of liberty, stretching back to the Magna Carta… [with] greater protection of human rights and private property' (Berg, 2012, np).

An outcome of Empire was the adoption of English as the lingua franca, facilitated by Britain's growing economic, trade, maritime and military dominance from the late 18th century, with British victory at Trafalgar in 1805 and Waterloo in 1815, followed by a century of Pax Britannica. Its spread was also facilitated by British influence in overseas territories, as well as formal and informal empire-building in the 19th and 20th centuries (Mandel, 2005). With international hegemony passing to the English-speaking United States after the Second World War, the centrality of English as the language of trade and diplomacy continued. Moreover, the post-war international dominance in the popular culture of the United States, and to a lesser extent the UK, provides cultural or 'soft' power manifested through global media, Hollywood and other commercial and 'blockbuster' movies, television and the pop music industry, and fashion industries in Los Angeles, New York and London. The BBC radio World Service and TV BBC World News broadcasts are challenged in global markets by US-established English-speaking networks such as CNN and Fox. This, according to Parulekar and Kotkin (2012), generates a belief in a 'linguistic superiority'; and it has also tied economic advancement to the acquisition of the English language, seen dramatically in India's and other English-speaking nations' key role in call centres servicing the world, and digital technological development by immigrants in Silicon Valley and other industrial districts. The English-speaking nations themselves are leading destinations for international students and workers due to a claimed strong sense of 'cultural belief', a 'rich heritage of freedom' and openness (Howard, 2011: 5). Of the over 900,000 students that China sent abroad, around 630,000 went to the United States, Australia, the UK, Canada and New Zealand, in that order (UNESCO, 2019); albeit whether this was to seek freedom as Howard implies rather than seeking advancement through English language acquisition is moot.

The construction of an Anglosphere has been a project of political leaders and other elites, particularly as a counter to British Imperial decline, and in the face of rival powers not sharing democratic and other values. If an aggressive and heavily armed but economically vulnerable Soviet Union was a threat during the Cold War, an increasingly rich, powerful, militaristic, assertive and highly autocratic China is perhaps a greater long-term challenge. In his famous 'Iron Curtain' speech in 1946, Winston Churchill asserted a 'special relationship between the British Commonwealth and Empire and the United States', seeking to link together 'the abiding power of the British Empire and Commonwealth' with the moral, economic and military might of the United States (International Churchill Society, nd). More recently, former Australian Prime Minister John Howard noted:

> you can be bound together by treaties [or] common economic objectives, but… the bonds are never as close as between countries that have shared values. This is not a time to apologize for our particular identity, but rather to firmly and respectfully and robustly reassert it. (Howard, 2011: 3–5)

Moreover, politicians, non-governmental organisations (NGOs) and think tanks in multiple states made the construction of a united Commonwealth, albeit largely the former 'white dominions', a policy position. In March 2015, the Commonwealth Freedom of Movement Organisation (later renamed CANZUK International) was founded in Canada to lobby for free trade and free movement of labour/travel between the UK, Australia, New Zealand and Canada, mirroring similar arrangements existing in the Australia and New Zealand Closer Economic Relations Trade Agreement.

Tourism, the Anglosphere and the Commonwealth

Travel itself encourages the celebration and reassertion of Britishness in various ways, and can work to bind together the so-called Anglosphere and the Commonwealth. English-speaking tourists are more likely to visit cultural attractions in other English-speaking countries (Foo & Rossetto, 1998); albeit tourism elsewhere can foster national identity, with British tourism to Spanish Mallorca found to establish a sense of 'Britishness' and 'whiteness' through familiar drink, food, language and flags (Andrews, 2005). Moreover, British heritage celebrating former greatness and affirming national identity is itself a tourism attraction, attracting vast numbers both from within and outside Britain. This also reaffirms and to some extent creates a perhaps imagined Commonwealth identity. The National Maritime Museum at Greenwich celebrates the rise of Britannia to 'rule the waves', glorifying Horatio Nelson, James Cook, Ernest Shackleton and David Livingstone among others; many of whom (especially Cook) resonate with former colonies (Horne, 1984). The National Trust, private castles and houses, and numerous museums and art galleries, attract vast numbers, often with extensive collections devoted to (and taken from) former colonial possessions. The monuments celebrating Britain's role in the Second World War, with tunnels at Dover Castle and Churchill's bunkers, and monuments to the war dead in London and across the country, focus on the heroic sacrifice of Empire and Commonwealth. Visitors and immigrants to the UK are struck by the seeming obsession with the Second World War – particularly the Blitz, the Battle of Britain and Winston Churchill – with endless documentaries, highly successful feature films, TV dramas and tourist attractions. Constant references to British, and particularly London, places and monuments in literature, film and hugely successful British television and other popular media exported around the world make parts of Britain strangely – and often reassuringly – familiar to even first-time tourists. Britain's former possessions themselves are sites of nostalgia and tourism: from the great Indian cities of the Raj, colonial-era hotels in Sri Lanka, Singapore and Fiji and elsewhere, and the old cities of Penang and Singapore, to the colonial buildings, railways and monuments scattered across the world.

The head of the Commonwealth, Queen Elizabeth II and the monarchy itself plays an important part in this linking of former Empire, trade, Commonwealth and tourism, establishing a British brand and asserting a Commonwealth community (Balmer, 2011). Formal constitutional responsibilities combine with symbolic, religious and celebrity roles, which feed an almost mystical fascination with royalty itself. Royal castles, museums, pageantry and historical narratives are major tourist attractions (Long & Palmer, 2007). Members of the monarchy are global figures. Diana, Princess of Wales, and more recently Prince Harry and his Hollywood actress wife Meghan, are among the most well-known celebrities on the planet, with the minutiae of their lives followed closely in the world's media. Diana's marriage to Charles attracted television audiences estimated at 1 billion and the 2011 wedding of Prince William, second in line to the throne, attracted around 400 million. Over one-third of the UK population watched Prince Harry and Meghan Markle's wedding in 2018; 1 in 10 in the United States alone (Fitzgerald, 2018). Windsor Castle, where the wedding was held, experienced a 92% increase in visitors in the three months following the event (Mzezewa, 2019). Diana's funeral in 1997 attracted the largest British live TV audience on record, with claimed worldwide audiences of 2.5 billion, and with her cult-like memorials attracting tourists to this day (Oliphant, 2017). This linking of Britishness, monarchy and tourism connects international celebrity and membership of the Commonwealth. The annual contribution of the royals to the British economy is estimated to be £1.8 billion. This includes £550 million in tourism and an increase in trade of around £150 million due to royal diplomatic efforts (Chu, 2018).

The Queen as head of state of 16 counties and leader of the Commonwealth actively demonstrates a commitment to the Commonwealth. As a princess in 1947, she pledged devotion to 'the service of our great imperial family to which we all belong' (Yang, 2014: 89), and since the coronation in 1953 has visited all countries of the Commonwealth at least once, save Cameroon and Rwanda which joined in 1995 and 2009, respectively. Moreover, soft power diplomacy through visits by other royals to members of the Commonwealth, included tours by Charles and Diana soon after their marriage; and visits to the United States, where celebrity status combines with broader constitutional and other roles. Royals such as Prince Andrew have been active in supporting trade missions, particularly to the Middle East. In 2019, Prince Harry and his wife visited coup-prone Fiji, now a republic and a reinstated but formally twice-suspended member of the Commonwealth; at the same time, British diplomatic and aid efforts have been strengthened in the Pacific.

Contested Heritage and the Politics of Disengagement

Language and community can both include and exclude. The notion of a revitalised Commonwealth attracts claims of a focus on a 'white' and rich Commonwealth – what were the self-governing Dominions before the Second World War – and where the linking of tourism, heritage and Commonwealth sanitises a contested and bloody history of Empire (Namusoke, 2016). Critics such as Susen (2017: 70) see Brexit support in racial, or perhaps racist, terms with a desire for an 'independent Brexitania... based on a mono-culturally defined sense of pure, pristine and patriotic Britishness'. A broader post-Brexit engagement with a Commonwealth, seems then to be two different agendas: first, a renewed engagement with the former (and now rich) white dominions, tying into the notion of a more British-focused community; second, a more encompassing engagement with the entire Commonwealth, with a range of languages, ethnicities and religions; but bound together through shared values and history.

Critics of proponents of a revitalised English-speaking world see them as having an uncritical view of British history, with an obsession for past glories of Empire representing 'a kind of jingoistic and militaristic nostalgia for the nation's status as a global power' (Day & Lunn, 2003;

Lloyd, 2019: np; Mycock & Wellings, 2017). Indeed, this can be reflected in tourism. For Hirsch (2019):

> Tour operators, hotel and museum owners think that visitors want to hear something that conforms to their already comfortable worldview. It's hard enough to get people to think about these questions here in Britain. But if Britons are still being helped to avoid the truth of empire in the places where it happened, and where local people are still living with the consequences, then what hope have we got?

British museums have long been condemned for their 'heritage of plunder' of cultural and sacred artefacts from other societies and ethnic groups (Chamberlin, 1983).

The commodification of history as a tourist package is claimed to create 'dissonant heritage' (Tunbridge & Ashworth, 1996), where grand narratives of benign Empire subsume more nuanced and critical local history, and where particular histories of colonisers take precedence over those of the colonised or dominated. In the Caribbean, particularly the former British Bahamas, the tourism industry places an over-emphasis on historical links to British colonialism. This focus deflects the attention of tourists away from seeing or imagining the contemporary experiences of local people and their authentic lives (Palmer, 1994).

Contesting interpretations of history are seen in New Zealand's celebration of 250 years since Captain James Cook's arrival. In one small village, local Māori refused permission for the six vessel flotilla containing a replica of Cook's ship, the Endeavour, along with Māori and Tahitian ocean-going canoes, to dock. A statute in Gisborne where Cook first landed was vandalised and sprayed with the words: 'This is our land' and 'Thief *Pakeha* [white person]' (Russell, 2019), and Endeavour's visit in October 2019 was met with protesters waving Māori flags. Some countryside destinations in the UK are seen to be hostile to those perceived as un-British due to ethnic or religious identities, such as Commonwealth Muslim tourists and Black British descendants from the Commonwealth Caribbean (Stephenson, 2007; Stephenson & Ali, 2010). While moves are made to increase connections with the white Commonwealth, the 'Windrush Generation' of black Caribbean immigrants has faced deportation; despite living in the UK for a significant period (Gentleman, 2019).

Heritage tourism can hide these racial and hierarchical divides, it is claimed, particularly through grand hotels and other building scattered though former Empires (Peleggi, 2005). Robert's (2017: 41–47) sees efforts of the Yangon Heritage Trust to preserve British colonial architecture running 'the risk of erasing colonial abuses and masking entrenched urban inequalities' and undermining efforts to celebrate post-1945 history and Yangon's role as the 'capital of independent Burma'; with these buildings being symbols of 'British right to rule'. Raffles Hotel in Singapore capitalises on nostalgia, with the name used for a hospital, golf club, marina, shopping mall, business class on Singapore International Airline, schools and streets; all linking the 'superior quality of the product or service with connotations of exclusivity and upward mobility' (Henderson, 2004a: 121); and perhaps playing down the inequities, racial hierarchies and constraints of Empire. The reopening of the Grand Pacific Hotel in Suva, Fiji, perhaps plays a similar role. Built in early 20th-century Edwardian grandeur by the New Zealand Union Steam Ship Company, the hotel was near derelict for 20 years before its restoration and reopening in 2014. Its advertising makes direct reference to visits of former colonial elites, including Royal visits, and is seen by Cheer and Reeves (2015: 161) as replicating the racial and hierarchical divisions of colonial times, and the privileging of tourism – carried out in their words by people of leisure and 'nouveau riche' – over the demands of a developing state.

The Ambiguities of Commonwealth Tourism

Gopal (2019) claims that Britain has failed to critically engage with Empire and colonialism. British historical education can gloss over and/or ignore the dark sides of Empire (Milne, 2012). Only in recent decades have such atrocities as the suppression of the Kenyan Mau Mau rebellion been widely discussed and litigated successfully in British courts, leading to compensation for victims (Parry, 2016). However, the issue remains unremarked in the state-operated National Museums of Kenya (Munene, 2011). The British courts held the British government responsible for the 1948 Malaysian Batang Kali massacre of 24 suspected and unarmed communist sympathisers. Although there have been calls to fund a monument to the dead (Meikeng, 2013), the massacre receives little attention in Batang Kali, with the best sights listed as a temple, a waterfall and a traditional food shop.

Nonetheless, some academic literature can present tourism in the former Empire in a perhaps wholly and unambiguously negative light – simply nostalgia and myth-making – with smug tourists touring the world to feel good about themselves, and crimes of Empire erased from history (see Cheer & Reeves, 2015). This caricatures sophisticated tourists and sophisticated locals too, who may be adept at navigating complex, ambiguous and contesting narratives of history, and where tourism itself facilitates engagement with critiques of Empire. A restaurant in Alexandria, Egypt, has on its walls photos of the destruction caused by the Anglo-French 1882 bombardment. Former prisons across Australia, and particularly Tasmania's Port Arthur, provide an achingly stark reminder of the brutalities of British transportation of prisoners. The horrors, complexities and ambiguities of the intentional slave trade are seen first-hand by tourists visiting slave castles in Ghana and elsewhere, with the involvement of European empires in slavery actively facilitated by local elites (Shenoy, 2019). Liverpool's International Slavery Museum is located only metres from where 18th-century slave ships docked. Liverpool's World Museum (2019) actively 'interrogates and links historical and cultural themes' and the complex history of acquisition and Empire, with a recent exhibition on the Benin Bronzes seized in a punitive exhibition by Britain in 1897, entitled: 'What to do with your loot'. A joint project of the BBC and the British Museum sought to challenge negative representations of countries such as Germany, presenting a range of 'triumphs and tragedies' over the past six centuries, with the legitimacy, scope and vastness of the museum's collection allowing for this contextualisation and critical engagement (Wilson, 2010).

Some past transgressions – or claims or perceptions of such – are also being addressed. The Human Tissue Act of 2004[1] allows museums to return human remains less than 1000 years old, particularly 'if it appears to them to be appropriate to do so for any reason'. This has seen the Natural History Museum (London) returning Torres Strait Islander bones; the World Museum (Liverpool) returning a skull to representatives of the Ngarrindjeri people of South Australia; and the British Museum returning two cremation ash bundles to the Tasmanian Aboriginal Centre. Nine human bone fragments were transferred to Te Papa, New Zealand's national museum. This has not been a carte blanche return, with competing narratives of science, education, preservation and access (Wilding, 2019). The British Museum declined to hand over seven preserved Māori tattooed heads (in which there was a lively trade in early New Zealand), as well as skulls from the Torres Strait Island decorated for use in divination (Shariatmadari, 2019). The return of the Parthenon/Elgin Marbles to Greece, for which there is now a dedicated museum in Athens, remains a cause celebre.

That colonial buildings and other monuments are simply expressions of power and perhaps oppression, to be removed from the landscape, also caricatures a more complex relationship between history, reuse, repurposing and re-interpretation. Well-built colonial buildings can be both a practical economic resource and an aesthetic one, with significant financial value (Heintzelman & Altieri, 2013). Marvellous Melbourne with its grand Victorian buildings and parks, and the heritage areas of colonial Sydney, can be contrasted to the blasted and charmless 'modernised' centres of some other Australasian cities. Madras/Chennai, Bombay/Mumbai and New Delhi are grand cities partly because of their remarkable and celebrated colonial architecture. Singapore and Malaysia celebrate colonial districts to be conserved and marketed as attractions (Henderson, 2004b). Fiji's restored colonial-era five-star Grand Pacific Hotel is as much a symbol of a reinvigorated and resurgent Fiji after decades of strife, as it is the nostalgic colonial relic caricatured by Cheer and Reeves (2015), with its recent restoration funded by Fijian and PNG money, and the tourism and the employment created as useful ways of addressing development issues. New indigenous Fijian and Indo-Fijian elites subvert symbols of former colonial power by drinking and eating in the very bars and restaurants their ancestors may have been excluded from in the past, dealing with foreign elites now as supplicants. Yangon's elites seek to preserve the city's grand colonial buildings for their particular tastes and interests and perhaps because they represent one of the more remarkable and best preserved albeit crumbling collection of such buildings in existence; whatever sinister symbolism may be imputed to them by well-meaning Western academics (Roberts, 2017).

Conclusion

A post-Brexit world opens up new possibilities for a revitalised Commonwealth, providing new opportunities for trade, investment, labour flow, technology transfer and, importantly, for this collection, tourism. We note different strands in the literature in this Commonwealth re-engagement narrative: first, re-engagement with rich and historically white dominions or settler societies (perhaps plus the United States), with this Anglosphere and Anglophone 'Better British' community substituting for lost communities of the EU; second, a more encompassing vision of Commonwealth, linking members of different ethnicities, income and size across the world, through shared historical links and values. Neither are mutually exclusive, of course. But both smack of desperation in the face of undoubted costs of leaving the EU in terms of lost income (Head, 2016), lost influence and perhaps ease of tourism.

It is unlikely that even the vast Commonwealth will simply swop for the closer, and rich, EU, anytime soon. Australia, New Zealand and Canada have moved on from the pre-EEC days when they were tied more closely to the British economy, and sought markets and contacts elsewhere. Some members of the Commonwealth – notably India – are not overly enamoured of free trade and open economies. The days of a free trade Empire overseen by a benign Royal Navy are long gone, if they ever existed.

But saying the project is somewhat fanciful, is not to say that it is entirely so. Despite claims that such projects are an exercise solely in nostalgia, or even that they reflect an inherent racism of the Brexit project, it seems trite to simply dismiss the value of the Commonwealth on those grounds alone, or discount the value and interest that shared histories and cultures create. Britain will need friends where it can find them. The values that the Commonwealth adopts and promulgates such as liberal democracy and mutual respect – even if some members are hardly exemplars in their application – cannot simply be dismissed as sentimentalism or nostalgia. To many, they are inherently important, and perhaps now even more so with rising powers that do not share them. Nor is travel around the former Empire necessarily or simply a celebration and ritualised forgetting of the darker aspects of Empire. Travel might well narrow the mind, but it can also expose individuals to challenging and different views of history. Engaging with the Commonwealth and its history and diversity, particularly through tourism, may well be a project worth investigating.

Note

(1) This is an act for England, Wales, and Northern Ireland. Scotland has the Human Tissue Act 2006.

References

ABC (2019) Britain wants trade deal with Australia within months after leaving EU. See https://www.abc.net.au/news/2019-09-18/britain-seeks-post-brexit-australian-trade-deal-within-months/11525200 (accessed 10 November 2019).

Andrews, H. (2005) Feeling at home: Embodying Britishness in a Spanish charter tourist resort. *Tourist Studies* 5 (3), 247–266.

Ashkanasy, N.M., Trevor-Roberts, E. and Earnshaw, L. (2002) The Anglo cluster: Legacy of the British Empire. *Journal of World Business* 37 (1), 28–39.

Balmer, J.M.T. (2011) Corporate heritage identities, corporate heritage brands and the multiple heritage identities of the British Monarchy. *European Journal of Marketing* 45 (9/10), 1380–1398.

Bennett, J.C. (2007) *The Anglosphere Challenge: Why the English-Speaking Nations will Lead the Way in the Twenty-First Century*. Plymouth: Rowan and Littlefield.

Berg, C. (2012) There is something good in the Anglosphere. *Sydney Morning Herald*, 12 August. See https://www.smh.com.au/politics/federal/there-is-something-good-in-the-anglosphere-20120811-2414g.html (accessed 15 March 2014).

Business Times (2018) UK's May looks to Commonwealth as a new source of trade, 17 April. See https://webcache.googleusercontent.com/search?q=cache:ddvy2n5dlpgJ:https://www.businesstimes.com.sg/government-economy/uks-may-looks-to-commonwealth-as-new-source-of-trade+&cd=1&hl=en&ct=clnk&gl=my (accessed 15 March 2019).

Chamberlin, R. (1983) *Loot: The Heritage of Plunder*. London: Thames and Hudson.

Cheer, J.M. and Reeves, K.J. (2015) Colonial heritage and tourism: Ethnic landscape perspectives. *Journal of Heritage Tourism* 10 (2), 151–166.

Chu, B. (2018) Does the Royal family really make financial sense for the UK economy? *Independent*, 11 May. See https://www.independent.co.uk/news/business/analysis-and-features/royal-wedding-family-how-much-uk-economy-benefits-cost-meghan-markle-expense-a8345436.html (accessed 5 November 2019).

Commonwealth, The (nd) See https://thecommonwealth.org/ (accessed 28 October 2019).

Day, A. and Lunn, K. (2003) British maritime heritage: Carried along by the currents? *International Journal of Heritage Studies* 9 (4), 289–305.

Dilley, A. (2016) The Commonwealth is not an alternative to the EU for Britain. *The Conversation*, 8 April. See https://theconversation.com/the-commonwealth-is-not-an-alternative-to-the-eu-for-britain-57009 (accessed 12 June 2019).

Fitzgerald, T. (2018) Royal wedding ratings: How many people watched Prince Harry wed Meghan Markle? *Forbes*, 21 May.

Foo, L.M. and Rossetto, A. (1998) *Cultural Tourism in Australia: Characteristics and Motivations*. Canberra: Bureau of Tourism Research; DCITA.

Gentleman, A. (2019) Chased into 'self-deportation': The most disturbing Windrush case so far. *Guardian*, 14 September. See https://www.theguardian.com/uk-news/2019/sep

/14/scale-misery-devastating-inside-story-reporting-windrush-scandal (accessed 12 November 2019).

Goldfinch, S. and Mein-Smith, P. (2006) Compulsory arbitration and the Australasian model of state development: Policy transfer, learning, and innovation. *Journal of Policy History* 18 (4), 419–445.

Gopal, P. (2019) *Insurgent Empire: Anticolonial Resistance and British Dissent*. London: Verso Books.

Grenade, W.C. (2016) Paradoxes of regionalism and democracy: Brexit's lessons for the commonwealth. *The Commonwealth Journal of International Affairs* 105 (5), 509–518.

Grenade, W.C. (2018) Paradoxes of regionalism and democracy: Brexit's lessons for the commonwealth. In P. Clegg (ed.) *Brexit and the Commonwealth: What Next?* (pp. 61–70). London: Routledge.

Head, S. (2016) The death of British business. *The New York Review of Books*, 18 October. See https://www.nybooks.com/daily/2016/10/18/brexit-death-of-british-business/ (accessed 16 November 2019).

Heintzelman, M.D. and Altieri, J.A. (2013) Historic preservation: Preserving value? *The Journal of Real Estate Finance and Economics* 46 (3), 543–563.

Henderson, J. (2004a) British colonial heritage in Malaysia and Singapore. In C.M. Hall and H. Tucker (eds) *Tourism and Postcolonialism: Contested Discourses, Identities and Representations* (pp. 113–125). London: Routledge.

Henderson, J. (2004b) Conserving colonial heritage: Raffles Hotel in Singapore. *International Journal of Heritage Studies* 7 (1), 7–24.

Hirsch, A. (2019) Teach British tourists the truth about empire, they can take it. *Guardian*, 20 June. See https://www.theguardian.com/commentisfree /2019/jun/20/british-tourists-empire-hotels-museums (accessed 20 June 2019).

Hofstede, G. (2001) *Culture's Consequences: Comparing Values, Behaviors, Institutions, and Organizations across Nations* (2nd edn). Thousand Oaks, CA: SAGE.

Horne, D. (1984) *The Great Museum: The Re-Presentation of History*. London: Pluto.

Howard, J. (2011) The Anglosphere and the Advance of Freedom. The Heritage Foundation, January 3, No. 1176. See https://www.heritage.org/report/the-anglosphere-and-the-advance-freedom (accessed 20 June 2019).

International Churchill Society (nd) The Sinews of Peace ('Iron Curtain Speech'). See https://winstonchurchill.org/resources/speeches/1946-1963-elder-statesman/the-sinews-of-peace/ (accessed 7 September 2019).

Lloyd, D. (2019) Debate: Is British history too fixated with the story of the world wars? In *History Extra*, June 24. See https://www.historyextra.com/period/first-world-war/debate-is-british-history-too-fixated-with-the-story-of-the-world-wars/ (accessed 29 September 2019).

Long, P.E. and Palmer, N.J. (eds) (2007) *Royal Tourism: Excursions around Monarchy*. Clevedon: Chanel View Publications.

Mandel, D. (2005) The 'secret' history of the Anglosphere. Institute of Public Affairs Review. *A Quarterly Review of Politics and Public Affairs* 57 (4), 31–32.

Meikeng, Y. (2013) Relatives want Batang Kali memorial. *Star*, 19 June. See https://www.thestar.com.my/news/nation/2013/06/19/relatives-want-batang-kali-memorial-families-hope-british-govt-will-honour-massacre-victims-and-apol#jS3ysgtgROfiz Lrg.99 (accessed 2 October 2019).

Milne, S. (2012) *The Revenge of History: The Battle for the Twenty-First Century*. London: Verso.

Munene, K. (2011) Museums in Kenya: Spaces for selecting, ordering and erasing memories of identity and nationhood. *Journal of African Studies* 70 (2), 224–245.

Mycock, A. and Wellings, B. (2017) The Anglosphere: Past, present and future. *British Academy Review* 31, 42–45.

Mzezewa, T. (2019) Getting on a plane for the Royal baby. *New York Times*, 11 April. See https://www.nytimes.com/2019/04/11/travel/meghan-markle-baby-sussex-travel.html (accessed 17 June 2019).

Namusoke, E. (2016) A divided family: Race, the Commonwealth and Brexit. *The Round Table* 105 (5), 463–476.

Oliphant, N. (2017) Princess Diana's funeral still most watched TV event in history 20 years on from death. *Express*, 6 September. See https://www.express.co.uk/news/royal/850523/Princess-Diana-death-funeral-Prince-Charles-Prince-William-Prince-Harry-TV-event (accessed 21 October 2019).

O'Neil, A. (2017) Australia and the 'Five Eyes' intelligence network: The perils of an asymmetric alliance. *Australia Journal of International Affairs* 71 (5), 529–543.

Palmer, C. (1994) Tourism and colonialism: The experience of the Bahamas. *Annals of Tourism Research* 21 (4), 792–811.

Parulekar, S. and Kotkin, J. (2012) The state of the Anglosphere: The decline of the English-speaking world has been greatly exaggerated. *City Journal*, Winter. See https://www.city-journal.org/html/state-anglosphere-13447.html (accessed 1 September 2019).

Parry, M. (2016) Uncovering the brutal truths about the British Empire. *Guardian*, 18 August. See https://www.theguardian.com/news/2016/aug/18/uncovering-truth-british-empire-caroline-elkins-mau-mau (accessed 21 August 2019).

Peleggi, M. (2005) Consuming colonial nostalgia: The monumentalisation of historic hotels in urban South-East Asia. *Asian Pacific Viewpoints* 46 (3), 255–265.

Roberts, J.L. (2017) Heritage-making and post-coloniality in Yangon, Myanmar. In H.H.M. Hsiao, H. Yew-Foong and P. Peycam (eds) *Citizens, Civil Society and Heritage-Making in Asia* (pp. 40–60). Singapore: ISEAS Publishing.

Russell, G. (2019) UK expresses 'regret' over Māori killings after Cook's arrival in New Zealand. *Guardian*, 7 December. See https://www.theguardian.com/world/2019/sep/17/hes-a-barbarian-maori-tribe-bans-replica-of-captain-cooks-ship-from-port (accessed 9 October 2019).

Shariatmadari, D. (2019) 'They're not property': The people who want their ancestors back from British museums. *The Guardian*, 23 April. See https://www.theguardian.com/culture/2019/apr/23/theyre-not-property-the-people-who-want-their-ancestors-back-from-british-museums (accessed 23 September 2019).

Shenoy, R. (2019) 'Willful (sic) amnesia': How Africans forgot – and remembered – their role in the slave trade. *The World*, 20 August. See https://www.pri.org/stories/2019-08-20/willful-amnesia-how-africans-forgot-and-remembered-their-role-slave-trade (accessed 1 December 2019).

Stephenson, M.L. (2007) The socio-political implications of rural racism and tourism experiences. In J. Tribe and D. Airey (eds) *New Directions, Critical Challenges and Fresh Applications in Tourism Research* (pp. 171–184). Oxford: Elsevier.

Stephenson, M.L. and Ali, N. (2010) Tourism and Islamophobia: Muslims in non-Muslim states. In N. Scott and J. Jafari (eds) *Tourism in the Muslim World* (pp. 235–251). Bingley: Emerald.

Susen, S. (2017) No exit from Brexit. In W. Outhwaite (ed.) *Brexit: Sociological Responses* (pp. 153–182). London: Anthem Press.

Sykes, O. (2018) Post-geography worlds, new dominions, left behind regions, and 'other' places: Unpacking some spatial imaginaries of the UK's 'Brexit' debate. *Space and Polity* 22 (2), 137–161.

Tunbridge, J. and Ashworth, G. (1996) *Dissonant Heritage: The Management of the Past as a Resource in Conflict*. Chichester: Wiley.

UNESCO (2019) Global flow of tertiary-level students. See http://uis.unesco.org/en/uis-student-flow (accessed 22 October 2019).

Ward, M. (2019) Statistics on UK trade with the Commonwealth. Briefing Paper, Number CBP 8282, 31 July. House of Commons Library, London.

Wilding, M. (2019) Museums grapple with rise in pleas for return of foreign treasures, *Guardian*, 18 February. See https://www.theguardian.com/uk-news/2019/feb/18/uk-museums-face-pressure-to-repatriate-foreign-items (accessed 23 September 2019).

Wilson, B. (2010) Neil McGregor on a history of the world in 100 objects. *Telegraph*, 15 January. See https://www.telegraph.co.uk/culture/tvandradio/6997778/Neil-McGregor-on-A-History-of-the-World-in-100-Objects.html (accessed 22 September 2019).

World Museum (2019) What to do with your loot: Renovating the Benin Displays at World Museum. Blog by Zachary, 25 October. See https://blog.liverpoolmuseums.org.uk/author/zacharyk/ (accessed 27 October 2019).

Yang, U.E. (2014) *Last Great Queen? Elizabeth 11, Mother of Leadership, Seen from the Crown*. Bloomington, IN: AuthorHouse UK Ltd.

11 Brexit and the UK Overseas Territories: Tourism and the Reconstitution of Core–Periphery Identity

Maria Amoamo

Introduction

Geographically spread across the world there are 14 United Kingdom Overseas Territories (UKOTs); most of which are islands. Following the outcome of the UK's referendum on 23 June 2016, Britain's decision to exit the European Union (EU) (Brexit) has major implications for the UKOTs in respect of trade, security, financial aid, access and mobility, national identity and economic development including tourism. This chapter discusses the theme of tourism and Brexit with a focus on the core–periphery relationship and reconstitution of identity as the UKOTs articulate their future relationships with the UK post-Brexit. The yet unknown outcome of Brexit (at the time of writing) poses both risks and opportunities for the UKOTs, and potential re-imaging/imagining of small islands as sites of reciprocal power projection in the process. The choice for Brexit has drawn attention to the complex and variegated quality of Britishness across the territories of the UK (Wincott *et al.*, 2017: 430). More globally, Brexit invites observation of the fundamental changes that concepts of sovereignty, self-determination, sense of belonging and the nation-state are currently undergoing (Cassidy *et al.*, 2018). In a tourism context, it marks a starting point for the ongoing (re)imagination of tourism and travel mobilities post-Brexit. Consequently, Brexit futures open up an opportunity to reflect on how spaces for new and emerging forms of solidarity and identity are created, reworked or closed (Anderson & Wilson, 2018: 293). As such, the majority of the UKOTs will undergo some form of sociopolitical change (i.e. realignment) to their small island communities.

Brexit and Identity Politics

> Brexit provides a clear illustration of identity politics; both are riven with complexity and contradiction. (Gardner, 2017: 14)

Since the UK referendum, a wide range of literature has emerged examining the social, political and economic consequences of Brexit (Ashcroft & Bevir, 2016; Calhoun, 2016; Gilmartin *et al.*, 2018; Hopkin, 2017). Depending on the outcome, key issues involve sovereignty and control ('hard' Brexit with few ties left between the EU and UK) and collaboration and trade ('soft' Brexit maintaining a closely integrated relationship with the EU). The impact of shifting political framings also raises new issues of a de-bordered EU, a re-nationalised UK centre (Cassidy *et al.*, 2018: 188–189) or the opening of new forms of non-sovereign relations. Bishop and Clegg (2018: 331) argue that Brexit leaves small states in Europe with an increasing risk of a Franco-German 'cooperative hegemony' once the counterbalancing effect of UK power is withdrawn. As such, small EU states may struggle to have their say on joint decision-making if they do not maintain and strengthen collaborative relations. Moreover, Brexit has exposed differences between people and places within the UK, changing political moods and forms of belonging, attachment and identification (Anderson & Wilson, 2018). Being 'British' now encompasses a far broader understanding of social difference, inequality, separateness from others, than the terms 'national identity' and 'national belonging' suggested in the past (Cassidy *et al.*, 2018: 190). Some argue that the terms on which the debate around the referendum have taken place are symptomatic of a Britain struggling to conceive of its place in the world post-Empire.

This chapter uses a postcolonial lens to situate issues of sovereignty, national identity and the nation-state to discuss tourism and subnational island jurisdictions (SNIJs). Tourism has an intimate relationship with postcolonialism in that many former colonies have found popularity as tourist destinations. Postcolonialism is concerned with cultural engagement and provides a critical perspective that draws attention to the power structures constructed and maintained via forms of discourse; often built around the concept of resistance. The term invites more historically grounded analysis into the core–periphery relationship that exists in the SNIJ context, yet affords speculation on emergent post-Brexit identities which I term 'revisionary core–periphery'. The latter is contingent on two paradoxical dimensions; the first being that 'peripheries' (i.e. UKOTs) can strengthen in the reformation of newly formed national identities and second, identities can be simultaneously ideologically powerful yet fragile in the circumstances that may follow Brexit (Gardner, 2017). Revision as such can work to either strengthen or weaken core–periphery relations; and a reliance on tourism will play a part. What is apparent in negotiating Brexit is that UKOTs are forging strong(er) collective identities, so ensuring their sociocultural, economic and environmental needs are met as small island tourism destinations.

Core–Periphery: Subnational Island Jurisdictions

The tourism literature has consistently emphasised the increasing importance of the visitor industry in small islands (Butler, 1993; Harrison, 2001; Pratt, 2015). Additionally, research on the distinct development of small island tourism-driven economies (SITES) (McElroy & Hamma, 2010) exists alongside a wide range of research examining broader sociocultural, political and environmental issues faced by small island states (Briguglio, 1995; Cashman *et al.*, 2012; Graci & Dodds, 2010; Scheyvens & Momsen, 2008). The core–periphery dynamic is exemplified by small island developing states (SIDS) and SNIJs with varying levels of autonomy that are yet (or likely to be) independent. The majority of SNIJs do not seek sovereign status nor do they wish to lose their autonomous powers (Baldacchino, 2010) and the benefits of attachment to what is typically a larger and richer metropolitan power.

Research has shown that SIDS exhibit better economic indicators than larger developing states (McElroy & Pearce, 2006), and SNIJs possess competitive advantages at developing core competencies and nurturing economic diversity (Grydehoj, 2011). Baldacchino (2011: 243) advocates for 'strategic flexibility' versus 'vulnerability' as an approach to the performance of small island economies based on (island) actors' ability to practice 'change management'. In this way, living and coping with change results in 'rapid response capability' (Bertram & Poirine, 2007: 333) at multiple levels and is both a default/reactive and strategic/proactive disposition to opportunity (Baldacchino, 2011: 243–244). Arguably, the ability to embrace and deal with change at multiple scales builds resilience.

As a collective, island states constitute a growing presence within global entities, for example the UN General Assembly, notably asserting their collective voice on issues such as climate change (Weaver, 2017). Associated entities such as the 44-member Alliance of Small and Island States (AOSIS) are examples of active island paradiplomacy – or the capacity to engage with the outside world through varying patterns of communication and representation (Bartmann, 2006). Similarly, groupings such as the Overseas Countries and Territories (OCTs) (which includes the UKOTs) hold associate status with the EU, which fosters economic and social development between the OCTs and with the EU community as a whole. Growing cooperation between the UKOTs and OCTs has brought greater political visibility to the international stage and a stronger voice for SNIJs in the EU notwithstanding dyadic asymmetry ultimately tied to a core–periphery relationship.

The concept core–periphery has a well-established pedigree within the tourism literature and has been applied in several ways. Common themes include dependency (Chaperon & Bramwell, 2013) (in the economic sense) and the more generalised argument that island tourism development tends to follow the centre–periphery model of development (Sharpley, 2012: 168). Due to their limited size, isolation, marginalisation and resource limitations, Graci and Maher (2018: 250) note that islands can face significant challenges in securing the sustainable development of tourism. Likewise, Connell (2018) argues that islands are poorly placed to achieve sustainability and development, where critical decisions are made at scales beyond them. Notably, in tourism research the general idea of dependency theory in core–periphery relations has been criticised as deterministic, with the core seen as inevitably determining (Chaperon & Bramwell, 2013). Here, the 'agency' of actors within the periphery and their collective capacity to manage their circumstances go unrecognised. SNIJs exhibit a complex and dynamic orientation to the world system entailing trust-based increments in the empowerment of SNIJs towards economically and socially responsible agency.

Notwithstanding the quantity and quality of extant research including the application of core–periphery thinking to tourism on/to islands, some authors have recently criticised that 'the subject matter needs greater and more detailed discussion' (see Butler, 2017). Targeted research will enable useful revision of the core–periphery relations unique to SNIJs and to outpost tourism in general. A 'revisionary core–periphery' approach will counter a long-standing core–periphery narrative which holds that small islands are geographically and economically marginal entities fated to spawn homogeneous tourist monocultures within the context of persistent external dependency (Weaver, 2017: 11). Brexit offers the opportunity for UKOTs to recast the core–periphery relationship, to strengthen their peripheral identities and so reposition within the local/global and geopolitical milieu of outpost tourism.

The British Overseas Territories: Postcolonial identity

> I firmly believe that it is vital for our Overseas Territories to be vibrant and flourishing communities proudly retaining aspects of their British identity and generating wider opportunities for their people. (John Penrose, Minister for Tourism and Heritage, Foreign & Commonwealth Office [FCO] 2012)

Like other SNIJs, the UKOTs span all oceanic basins and boast all manner of diversities of size, topography, climate, ecology, history, economy, politics and cultural identities. As fragments of empire, the UKOTs exhibit the indelible imprint of a colonial past, and distinct traces of colonialism remain in the lack of political sovereignty and economic fragility; yet, none could aptly be described as a colony in the simplistic vernacular understanding of the term (Aldrich & Connell, 1998: 3).

Geographical remoteness both distances and differentiates these 'dots on the map' – a form of spatial erasure reinforced by descriptions of islands as tiny, remote and isolated. Erasure is inextricably tied to the imbalance of power and the scrutiny of processes of social negotiation and identity between core–periphery evidenced in the well-documented 'ad hoc' approach to administering the territories characteristic of past 'pragmatic' British colonial rule (Clegg & Gold, 2011: 2). Such island outposts were seen as something of an 'obligation that they (the UK) would shoulder, though without great enthusiasm or sense of national purpose' (Aldrich & Connell, 1998: 3). And Britain was not alone in maintaining such core–periphery entities; France, Denmark, Spain, the Netherlands, the United States, Australia and New Zealand all administer 'territories' or 'commonwealths' in some form or fashion.

The 14 'dependencies' (11 are permanently populated) of Britain include Gibraltar, the British Indian Ocean Territory (BIOT), South Georgia and the South Sandwich Islands, the Falklands, Pitcairn Island in the Pacific, St Helena (and its dependencies, Tristan da Cunha and Ascension); and in the Caribbean, Bermuda, the British Virgin Islands (BVI), Anguilla, Montserrat, the Cayman Islands and the Turks and Caicos Islands. The populated territories' main industries are tourism, financial services and fisheries and agriculture with the majority of territories relying heavily on one or two economic sectors. The financial services industry is one of the main contributors to the economies of Bermuda, the Cayman Islands, the British Virgin Islands and Gibraltar, while tourism is a significant part of the economy for the Caribbean islands of Anguilla, Montserrat and Turks and Caicos Islands and remote Pitcairn. Agriculture and fishing are significant sectors for the Falkland Islands, St Helena and Tristan da Cunha. The territories have vital interests in continuing to develop traditional economic sectors, but recognise the importance of continuing to diversify their economies. Several UKOTs such as Montserrat, Anguilla, St Helena and Pitcairn Island are not economically self-sufficient and depend on British and EU aid. Such differences affect the territories' current relations with the UK and affect the identity discourses of each (Harmer *et al.*, 2015).

In making a case for national identity, Benedict Anderson's (1991) work on the 'imagined community' helped inscribe the idea of national boundaries and differentiation. Cultural geographers have extended Anderson's insights to suggest that economics and geography influence expressions of national identity that undercut notions of national homogeneity; in particular relating to island spaces. Case study research on UKOTs (see Harmer *et al.*, 2015: 512) reveals that material factors and processes often express Overseas Territories' (OTs) identity formulations. Here, expressions of identity are influenced by issues of marginality, peripherality and otherness in relation to the metropolitan world, often described in terms of economic aid or geographical distance (Hall,

1990: 228). For example, OTs such as the Turks and Caicos Islands, the Cayman Islands, Bermuda and BVI are havens for offshore finance, insurance and investments, as well as tourism. The geographical separation and economic self-sufficiency of these islands are positive factors in their identity construction. In contrast, remote St Helena faces extreme problems of access and dependency on British aid, thus economic terminology or 'geographical analogy' is often used to call for greater recognition from the metropole (Harmer *et al.*, 2015: 513). St Helenians are described as being 'more British than the British'; prone to using a discourse that geographically places them within the geopolitical outline of the British Isles that draws on emotional language. For example, 'loyalty' (to the monarch) or 'patriotism' (St Helena British Armed Forces played a part in the Falklands War in 1982) to describe their link with the UK. Geopolitically, the 1982 invasion of the Falkland Islands cemented the relationship between British nationalism and the remote South Atlantic Island. McConnell and Dittmer (2018: 148) argue 'it was no longer necessary for the Islanders to argue for their Britishness: it was taken for granted, sealed in the blood of the British military'. So, it could be said that in some cases the OTs' relationship with the UK is shaped by an understanding of how they see themselves *in relation to* geographical, political, economic and cultural factors.

In the current environment, the wider possible impacts of Brexit for the UKOTs (apart from the risks posed to the security of Gibraltar and the Falkland Islands) was barely considered by the metropole leading up to and during the referendum period. Notwithstanding criticism of its past administrative oversight of the OTs, the past two decades has seen an attempt by the UK to re-engage with the territories through White Papers such as *Partnership for Progress and Prosperity* (1999) and *The Overseas Territories: Security, Success and Sustainability* (2012). The 1999 White Paper is notable for the UK government's reform related to the change in classification from 'Dependent Territory' to 'Overseas Territory' in the belief that it better reflected the nature of a postcolonial partnership (Clegg & Gold, 2011: 6). Empire and identity aside, a reworking of the core–periphery state assemblage is evident in the dynamism of political change: the current catalyst for rapid change being Brexit. Several distinct formations or multiplicities emerge being either political subordinate (periphery) to the UK (core) or as independent diplomatic actors with revisionary core–periphery capacity. Albeit the impact of Brexit will have different outcomes for the OTs, it is the *process* towards a post-Brexit environment that will likely enable more self-determination and with that, some reshaping of the identities of the OTs. The following section examines OTs' response and transformation from liminal/peripheral political subjectivities to one of *communitas* (McConnell, 2017) or shared equality through enforced engagement with the core *vis-à-vis* Brexit.

Realigning core–periphery relations: Brexit and the UKOTs

McConnell and Dittmer (2018: 140) argue that diplomatic encounters – operating in 'power spaces' – produce new potential for political change that is both dynamic and contingent. Turning attention to the oft-overlooked diplomatic articulations of the UKOTs, these authors contend that such entities can take on several distinct formations and relations that rework the state assemblage as 'becoming together' in ways that are productive of multiple political subjectivities. A 'revisionary core–periphery' shift sheds insight into the nature and dynamics of political subjectivity, practice and discourse – including the processes by which geopolitical categories and actors are produced (McConnell, 2017: 150).

Although historically, the UKOTs closest relationship has been with the UK, in recent years links with the EU have grown because of the Overseas Association Decision (OAD) adopted by the EU in 2013 from which the territories receive development assistance in the form of funding for social, cultural, economic and environmental initiatives (Clegg, 2016a). In the period leading up to the UK 2016 referendum, several MPs raised concerns regarding the lack of representation of OTs in the Brexit process and the potential risk to their position of strength as funding recipients in the EU OAD; a point reiterated by Tristan da Cunha's representative who stated:

> The risk that the OTs will simply be forgotten or side-lined as too complicated, too distant, and in some cases too small, for British politicians to be devoting significant share of mind to. (Megaw, 2017)

But since 2017, a process of strengthened engagement is evident in the creation of the UK-Overseas Territories Joint Ministerial Council on European Negotiations (JMC-OT EN) in addition to the annual November JMC[1] forum established in 2012 with a clear mandate to review and implement the shared strategy and commitments of the 2012 White Paper. The JMC also provides a forum to discuss views on politics, constitutional issues, security and the economic and social development of the OTs. Furthermore, with the UK's departure from the EU post-Brexit, the opportunity to strengthen UK/OT partnerships will become more important. That said, the conditions of Brexit pose risks and vulnerabilities for the OTs including the political geographies of new and growing forms of territorial fragmentation and uneven development.

In July 2017, the House of Lords (HOL) EU Select Committee heard evidence from representatives of the OTs (other than Gibraltar) on the impact of Brexit upon their territories. Several key issues were raised by the OTs (Table 11.1) including access to EU funding; trade; financial services; freedom of movement; relations with the EU and neighbouring territories; tourism and opportunities arising from Brexit. In particular funding via the European Development Fund (EDF) linked to the new

Table 11.1 Brexit and implications for the UKOTs

Brexit issues	Overseas Territory	Quote(s) from UKOT representatives to HOL EU Select Committee
Access to EU funding	Anguilla Tristan da Cunha St Helena	'the support from the EU from EDF funding accounted for some 36% of our capital budget in 2016... this is critical'. 'The EDF has taken a healthily multiannual view of infrastructure investment. That is our biggest concern'. 'Capital aid from the EU had allowed significant improvements to the island, including cliff stabilisation, wharf widening, customs and freight terminals, and rebuilding of roads'.
Trade	Montserrat Falkland Islands	'Our main export is the sand and aggregate that comes from the volcano... shipped to neighbouring islands or used for products that we can export. When we export to the French or Dutch territories, we enjoy tariff-free arrangements, which help us to be competitive. We dearly need to protect that one main export that we have'. 'We produce wool and meat as export crops. Currently, wool has no quote or tariff attached wherever it is sold. Our meat is slaughtered to EU standards... post-Brexit tariffs or taxes could apply'.
Financial services	Turks and Caicos British Virgin Islands (BVI)	'Funds of EDF 11 are assisting us in making our destination a bit more attractive and attracting foreign direct investment. That is important with a small population'. 'Financial services are very significant to the economy of the BVI... with Brexit, we have to be sure that the (supportive role) of the British Government will continue'.
Freedom of movement	Pitcairn	'We are an isolated island. Our nearest neighbour is French Polynesia. The southernmost part of that is 300 nautical miles away at Mangareva. It is not an international port. By agreement with our friends in French Polynesia and France, we use that as an international port. Will we still be allowed to do so? If not, we have serious problems in terms of our increasing isolation'.
Relations with other OCTs	Anguilla Montserrat	'The border relationship of Anguilla with Two EU member states is critical to our development... because of its relatively small population of 15,000 compared to a population of 100,000 in French and Dutch St Martin, Anguilla's people were reliant on their neighbours for many social services'. 'We have a population of 5,000 and therefore we cannot attract the specialist care that is needed. Oftentimes it is urgent medical care that has to be evacuated to Guadeloupe or neighbouring islands'.
Tourism	BVI Montserrat Anguilla	'we benefit from the EU's horizontal programmes, particularly when it comes to biodiversity... we have a project going on in the most tourist-oriented part of the island, where the beach and reef has been damaged by run-off'. 'Is seeking to use the current EDF tranche to develop a port to enhance Montserrat's tourism industry'. 'in its primary industry of tourism, 90% of visitors to Anguilla used St Martin as a hub... Many (Anguillans) go to St Martin to work in education, in the hotel sector, in construction and other areas of work'.

OAD has been a significant contributor to the annual budgets of the OTs. Total bilateral funding for the UKOTs (excluding Gibraltar) via the eleventh EDF (2014–2020) period is €76.8 million while regional funding is worth another €100 million, all of which focus on economic diversification, climate change mitigation and sustainable energy (Bishop & Clegg, 2018: 336). EU funding has allowed islands like St Helena and Montserrat to add or improve telecommunications connectivity, promote sustainable energy and reduce the risk from climate change and disasters. Several of the OTs also benefit from environmental programmes in the initiative BEST (Biodiversity and Ecosystem Services in Territories of Europe Overseas). Such funding has enabled Montserrat to develop a port to enhance their tourism industry; BVI to protect its rich biodiversity in a project that repaired damaged beach and reef in the most tourist-oriented part of the island; and Pitcairn Island to build infrastructure for visiting cruise ships.

The concerns raised by the OTs in Table 11.1 mark a critical junction in the core–periphery relationship in which the OTs must ensure that their interests are not neglected or actively marginalised (Clegg, 2016b). That said, opportunities arising from Brexit reflect changes in the traditional core–periphery relationship that enable more self-determination and with that, some reshaping of SNIJ identities.

Opportunities arising from Brexit

Notwithstanding the aforementioned concerns, particularly over trade and EU funding, some of the OTs see opportunities from Brexit. Two primary points are expansion into new markets and strengthening ties with regional and other international collectives. An example for the Caribbean islands is strengthening links with groups such as CARIFORUM (a body comprising Caribbean ACP states [African, Caribbean and Pacific Group] established in 1992) whose main objective is to promote economic integration and cooperation among its members (Griffith, 2002). ACP countries receive EDF funding from which the UKOTs also benefit. However, there is risk in amalgamating with larger groups like ACP resulting in the loss of a distinctive position as a collective (UKOTs). Brexit has also been a catalyst for encouraging a recommitment and adjustment of CARICOM (the 15-nation Caribbean community) of which the UKOTs are a part (Lewis, 2016). This community of small island states seeks, through regional solidarity, to boost their economies and mitigate their weak political influence in the global arena. Here, opportunities arise in bringing a new orientation to CARICOM's relationship with the UK while concurrently seeking new development possibilities for strategic alliances with other centres of influence such as the Community of Latin American and Caribbean States (CELAC). As members of the Organisation of Eastern Caribbean States and the Caribbean community, BVI and Anguilla (associate members) and Montserrat (full member) want to deepen these links, thus maintaining and potentially benefiting from the existing relationship these organisations have with the EU (Clegg, 2018: 161). Such regional economic partnership agreements (EPAs) could extend to other OTs such as St Helena, Tristan da Cunha and Ascension if endorsed by the UK, albeit with varying levels of local authority/autonomy depending on each OT's economic situation (Hendry & Dickson, 2011). Other territories like Pitcairn are keen to nurture and maintain strong links with French Polynesia while Anguilla is exploring the option of being part of the European Grouping of Territorial Cooperation. There is also support from other OCT member states for the UKOTs to remain associated with the EU with the suggestion that the UK contribute to EDF funds on a pay-as-you-go basis (Clegg, 2018). Autonomous arrangements already exist between the EU and UKOTs such as Bermuda, considered a 'third country' by the EU because of its link to the EU regulatory framework for the insurance market. According to Bermuda's political representatives, such arrangements are self-determining in respect of SNIJ autonomy and identity.

From another vantage point, the UKOTs provide geo-strategically dispersed locations that contribute to both biodiversity and maritime territory (Stoll-Davey, 2017). It is estimated that over 90% of the UK's biodiversity is in its OTs, providing a range of goods and services such as fishing, tourism and renewable energy supply. Moreover, the UKOTs also hold a vast range of endemic species, sensitive ecosystems and threatened species. Notwithstanding, biodiversity has underpinned generational sustainable livelihoods for OT inhabitants, and provides the potential to raise living standards and employment, especially through sustainable tourism. The European Commission (2006: 7) has highlighted 'the international importance for biodiversity' of the UK/OCTs, calling for a policy that recognises the need to balance both conservation and the development of natural areas. This aligns with the UK White Paper's focus on tourism as a major part of the economy of the territories – ensuring that growth in the UKOTs is 'sustainable, green and beneficial for their inhabitants' (Foreign & Commonwealth Office, 2012: 54–55). Stoll-Davey (2017) suggests that UKOTs may also be 'assets' of the EU post-Brexit in terms of the remarkable biodiversity and wealth of the territories' marine ecosystems, acting as locations for experimentation to combat the effects of climate change as well as scientific portals for their geographical areas. Albeit one of the less likely OTs to be independent, Pitcairn Island exhibits the ability to deal with situational change and, through tourism, is attempting to limit dependency and address the historical (im)balance of core–periphery relations.

Pitcairn Island Case Study

Geographically isolated and accessible only by ship, Pitcairn is one of the most isolated and smallest SNIJs in the world. Maritime enthusiasts and historians will recall Pitcairn as the refuge of the *Bounty* mutineers who settled the island with their Tahitian counterparts in 1790. The Pitcairn Island Group comprises four islands: Pitcairn, Ducie, Henderson and Oeno, with Pitcairn being the only inhabited island. The island became part of the British Empire in 1838; however, for much of its existence, British central administration was marginal, contributing in part to the formation of a somewhat insular society with its own laws and mores (Amoamo, 2017). In recent times, core–periphery relations have been strained due to the well-publicised UK investigation of historical sexual abuse resulting in criminal charges in 2004 and the subsequent jailing of several Pitcairn men. Notwithstanding negative media attention, the events divided the community, raised questions about the application of British law on Pitcairn (Lewis, 2009) and drew criticism of Britain's past neglect and now dominant and overbearing metropolitan authority on a small and vulnerable community (Trenwith, 2003).

Since then, Pitcairn has undergone a process of political, economic and social realignment that facilitates the re-imaging of island place and identity through 'revisionary core–periphery' relations. A programme of strengthened cooperation between the UK and Pitcairn has been agreed, with the adoption of generalised reciprocity (Putnam, 2000) rearticulated through a discourse of intentional agency, empowerment, self-determination, good governance and a shared vision of development. Today, tourism, primarily based on Pitcairn's *Bounty* heritage, offers the islanders the only viable means of economic growth and the prospect of a sustainable future. As such, tourism denotes a revisionary core–periphery realignment based on intentional agency with a view to 'change something to a different position or state' (MacLeod, 2013: 77).

Economic development and tourism on Pitcairn

Pitcairn is best described as a nano economy. There is limited economic activity including a tourism market (mostly based on homestays, visiting yachts and passing cruise ships), craft sales, production and sale of honey, basic agriculture and fishing. Off-island economic activities consist of the international sale of postage stamps and commemorative coins and internet domain registration (Department for International Development [DFID], 2017). Since 2004, Pitcairn has required UK development assistance with up to 95% of its budget requirements provided by the DFID. Ageing demographics (less than 50 people) coupled with a decline in the economically active members of the population mean there is little prospect of the island attaining self-sufficiency. That said, the Pitcairn community is determined to remain on the island and create a 'way of life' for future generations and potential new immigrants. Predictions that 'the end is nigh' (Connell, 1988) have been rife for decades, but a promising economic pathway has been forged through tourism development based on Pitcairn's natural ecology, *Bounty* heritage and unique isolation.

The past few years have seen substantial diversification of tourism-related activities funded by the DFID and the EU (Table 11.2) alongside

Table 11.2 Pitcairn EU-funded projects with tourism-related benefits

Renewable Energy Project	In 2017 a renewable energy feasibility study was undertaken on Pitcairn funded by the OCTA Innovation Programme of €5 million for technical assistance to meet the challenge 'Innovation for sustainable islands' growth'. The implementation of a renewable energy project would also be directly linked to Pitcairn's effort to re-populate the island and to diversify its economy (von Gesseneck & Zieroth, 2017: 38). As the only Pacific Island without a contribution of renewable energy to its electricity supply Pitcairn is solely reliant on diesel fuel for its electricity shipped from New Zealand. A proposal to replace 90% of the diesel fuel currently used is estimated at NZ$1,100,800 based on using a solar diesel hybrid system similar to that used on the Pacific atolls of Tokelau. From the perspective of tourism development the conversion from diesel to renewable energy is conducive to the SDTMP strategy of eco-friendly, blue-green image of Pitcairn which, together with the MPA and UNESCO site of Henderson, it is hoped such a power system would help to attract eco-tourism.
Glass Recycling Project	As part of the EU-funded sustainable development project INTEGRE – designed for and implemented by the four Pacific OCTs in 2017 Pitcairn was successful in its submission to projects benefiting the environment with a glass recycling plan. Coordination with a glass works expert from the Dutch Caribbean Island of Saba resulted in the joint visit to Pitcairn in August 2017 with INTEGRE representative to work with the island community and the Pitcairn Island Council's Division of Environment, Conservation & Natural Resources. As skilled artisans in their own right, adaptation to new ideas is an inherent part of Pitcairn culture, who have practiced the habit of recycling since long before the 'green' movement became a thing (The Pitcairn Miscellany, August 2017). The result of this project enables Pitkerners' to expand an already diverse range of curio products to sell to visiting (cruise) tourists.

a strong focus on biodiversity, marine and environmental conservation with the assistance of groups such as US-based Pew Charitable Trusts, National Geographic and the Royal Society for the Protection of Birds (RSPB). Environmental projects include a Sustainable Fisheries Management Plan (2012–2015), EU Overseas Territories Sustainable Energy Roadmap, rat eradication and restoration on Henderson Island (UNESCO World Heritage site), combating soil erosion (INTEGRE), water and waste management, and in 2015 the designation of a marine protected area (MPA) in the Pitcairn waters (some 800,000 km^2 of EEZ). In particular, the designation of the MPA and the associated positive media attention – after much lobbying by the Pitcairn community and non-governmental organisations (NGOs) – reinvested a sense of social identity, pride and ownership in the island's cultural landscape notwithstanding the potential economic benefits through ecotourism.

Notably, until 2011, Pitcairn was not marketed internationally as a tourist destination and its tourism infrastructure was minimal. A comprehensive Sustainable Tourism Development Master Plan 2015–2019 (STDMP) with a phased and managed approach to tourism development based on strategies of 'blue and green' is now in place. The latter enables sustainable environmental management, cultural enrichment, longevity and economic well-being alongside a policy to manage tourist capacity in tandem with infrastructure development and risk mitigation (Pitcairn Island Tourism Department, 2015: 2). Outsourcing expertise has enabled Pitcairn to manage capacity issues while building trade relations within the South Pacific and Australasia region. This includes strengthened media exposure and increased digital engagement with targeted online customers, with key tourism segments such as birding, diving and cruising given special focus and representation in the UK and the United States (Amoamo, 2017). Currently, Pitcairn is also exploring astro-tourism with the possibility of attaining Dark Sky Sanctuary (DSS) status with the International Dark-Sky Association.

With up to 16 cruise ships currently stopping at Pitcairn per year, cruise tourism is an integral part of Pitkerners' livelihoods with the opportunity to sell their crafts and souvenirs. As one of the fastest-growing sectors of the tourism industry in the South Pacific, cruise ship tourism is an attractive development option for small islands like Pitcairn, particularly those that lack the facilities to cater to large numbers of overnight tourists. That said, the rapid growth of the cruise industry has also raised awareness of its impacts on the natural environment, local communities and economies. Such controversies over the cruise sector's deficiencies are magnified in small island developing contexts with less diversified economies unable to compete with on-board mechanisms to capture passenger revenue or direct spending to allied firms (Cheer, 2017). Pitcairn is classified as a 'cruise-by' destination with only a small percentage of passengers landing on the island, a factor that safeguards environmental protection of the island (e.g. of 16 cruise ships that visited in 2014, only 4 landed passengers). To expand this sector of tourism, Pitcairn has joined the South Pacific Cruise Alliance group to attend overseas trade/marketing events such as the annual Seatrade Cruise Global convention (held in Fort Lauderdale USA in 2018). Pitcairn's Tourism Travel coordinator notes the benefits of collectiveness for the island: 'Working together is the way of the Pacific... our collaboration provides the entire team the opportunity to promote the South Pacific region and their individual islands to the international cruise company sector' (Pitcairn Miscellany, 2018: 9). There are also plans to develop a Pitcairn Island Eco Tourism Lodge over the next five years to provide more contemporary accommodation options for short-term visitors (Pitcairn Island Tourism Department, 2015: 10). The island's workforce has built infrastructure including an alternative harbour for visiting cruise ships, a cultural centre, a library and tourism facilities in the building that once housed the Pitcairn jail (and built by the islanders themselves).

As one of the smallest and most isolated UKOTs heavily reliant on UK aid, Pitcairn illustrates the resourcefulness, adaptation and capacity for economic resilience through revisionary approaches to the conventional demarcation of peripheries as disadvantaged relative to the core. Peripherality constitutes at least as much an opportunity as a threat when considered from a geopolitical tourism perspective and the re-imagination of tourism and travel mobilities post-Brexit. In the Brexit environment, Pitcairn has capitalised on the collective strength of other UKOTs, contributing to political dialogue and asserting its Pacific voice. Brexit arguably facilitates certain advantage and 'power' in the core–periphery relationship.

Conclusion

Examining Brexit in performative terms, Adler-Nissen *et al.* (2017: 582) argue that Brexit helps construct other national identities and subject positions in the milieu of bilateral and multilateral relations and the global economy. In an economic sense, Brexit has 'reshaped the ordering principles of the global economy' bringing together two largely antithetical positions: liberalism and economic nationalism. The latter is often associated with retreats from globalisation (e.g. 'taking back control' as in the election of Donald Trump to the US presidency) while the former is intrinsically linked to the growth of globalisation. Brexit constitutes the promise of a different future, and a different world order (Adler-Nissen *et al.*, 2017: 580). Thus, Brexit offers the UKOTs new paths of influence and opportunities to secure their own interests and development by building broader relationships and networks, regionally and globally.

Arguably, Brexit presents the OTs with a complex and competing set of challenges in achieving economic growth (for many through tourism development), promoting cultural identity and strengthening autonomy. Although the UKOTs have progressed to a measure of self-reliance and regional economic cooperation, the yet uncertain outcome of Brexit carries an element of vulnerability. Brexit likely entails cessation of EU funding provided to UKOTs to support tourism development and biodiversity. That places additional economic burden on the parent UK state. The UK government may then seek to tighten the supply of funds and resources to its peripheral territories. However, UKOTs have shown the ability to influence diplomatic practices through rapid response and to operate collectively to safeguard their interests. The Pitcairn case study expands the idea of tourism as a substantial factor in the development of resilience through economic diversity and core–periphery realignment.

The UKOTs' response to Brexit through 'revisionary core–periphery' relations can be seen as a move towards 'identity re-engineering' in the sense of bottom-up versus top-down strategic management (Jackson, 1996: 36). In an era of uncertain geopolitics and upheaval, revisioning is likely the default option in the face of the 'eclipse of government' (Jordan *et al.*, 2005). Brexit marks a critical junction in the core–periphery relationship in which proactive moves by the UKOTs to seek alliances, strengthen regional and global networks and pursue self-determining strategies are examples of revisionary core–periphery relations and transformation from liminal/peripheral political subjectivities to one of *communitas* (McConnell, 2017). Although the territories are vastly different in geographic and economic terms, they share the need for engagement with the core *vis-à-vis* Brexit. In terms of the core–periphery relationships, we might say that Brexit provides opportunity for OTs to realign the power structures constructed and maintained via forms of postcolonial discourse. As Brexit unfolds, a new paradigm shift from SNIJ dependency to opportunity contributes to a more nuanced understanding of tourism as a progressively geopolitical practice.

Note

(1) The JMC is an annual forum involving UK ministers and UKOTs' governments and was first convened in 2012.

References

Adler-Nissen, R., Galpin, C. and Rosamond, B. (2017) Performing Brexit: How a post-Brexit world is imagined outside the United Kingdom. *The British Journal of Politics and International Relations* 19 (3), 573–591.
Aldrich, R. and Connell, J. (1998) *The Last Colonies*. Cambridge: Cambridge University Press.
Amoamo, M. (2017) Re-imaging Pitcairn Island examining dualities of conflict and collaboration between island/metropole through Tourism. *Shima: The International Journal of Research into Island Cultures* 11 (1), 80–101.
Anderson, B. (1991) *Imagined Communities*. London: Verso.
Anderson, B. and Wilson, H.F. (2018) Everyday Brexits. *Area* 50, 291–295.
Ashcroft, R. and Bevir, M. (2016) Pluralism, national identity and citizenship: Britain after Brexit. *The Political Quarterly* 87 (3), 355–359.
Baldacchino, G. (2010) *Island Enclaves: Offshoring Strategies, Creative Governance, and Subnational Island Jurisdictions*. Montreal and Kingston: McGill-Queen's University Press.
Baldacchino, G. (2011) Surfers of the ocean waves: Change management, intersectoral migration and the economic development of small island states. *Asia Pacific Viewpoint* 52 (3), 236–246.
Bartmann, B. (2006) In or out: Sub-national island jurisdictions and the antechamber of paradiplomacy. *The Round Table* 95 (386), 541–560.
Bertram, G. and Poirine, B. (2007) Island political economy. In G. Baldacchino (ed.) *A World of Islands: An Island Studies Reader* (pp. 323–378). Charlottetown: Institute of Island Studies, Luqa and Malta: Agenda Academic.
Bishop, M.L. and Clegg, P. (2018) Brexit: Challenges and opportunities for small countries and territories. *The Round Table* 107 (3), 329–339.
Briguglio, L. (1995) Small island developing states and their economic vulnerabilities. *World Development* 23 (9), 1615–1632.
Butler, R. (1993) Tourism development in small islands: Past influences and future directions. In D.G. Lockhart, D. Drakakis-Smith and J.A. Schembri (eds) *The Development Process in Small Island States* (pp. 71–91). London: Routledge.
Butler, R. (ed.) (2017) *Tourism and Resilience*. Wallingford: CABI.
Calhoun, C. (2016) Brexit is a mutiny against the cosmopolitan elite. *New Perspectives Quarterly* 33, 50–58.
Cashman, A., Cumberbatch, J. and Moore, W. (2012) The effects of climate change on tourism in small states: Evidence from the Barbados case. *Tourism Review* 67 (3), 17–29.
Cassidy, K., Innocenti, P. and Bürkner, H.J. (2018) Brexit and new autochthonic politics of belonging. *Space and Polity* 22 (2), 188–204.

Chaperon, S. and Bramwell, B. (2013) Dependency and agency in peripheral tourism development. *Annals of Tourism Research* 40 (1), 132–154.

Cheer, J.M. (2017) Cruise tourism in a remote small island: High yield and low impact? In R. Dowling and C. Weeden (eds) *Cruise Ship Tourism* (2nd edn, pp. 408–423). Wallingford: CABI.

Clegg, P. (2016a) *The United Kingdom Overseas Territories and the European Union: Benefits and Prospects*. London: UKOTA. See www.ukota.org/manage/wp.../UKOTA-Final-Report-on-Benefits-of-EU-to-UKOTs.pdf (accessed 8 September 2017).

Clegg, P. (2016b) Brexit and the Overseas Territories: Repercussions for the periphery. *The Round Table* 105 (5), 543–555.

Clegg, P. (2018) The United Kingdom and its Overseas Territories: No longer a 'benevolent patron'? *Small States and Territories* 1 (2), 149–168.

Clegg, P. and Gold, P. (2011) The UK Overseas Territories: A decade of progress and prosperity. *Commonwealth & Comparative Politics* 49 (1), 115–135.

Connell, J. (1988) The end ever nigh: Contemporary population change on Pitcairn Island. *GeoJournal* 16 (2), 193–200.

Connell, J. (2018) Islands: Balancing development and sustainability? *Environmental Conservation* 1–14. See www.cambridge.org/core (accessed 25 May 2018).

Department for International Development (DFID) (2017) Business Case Pitcairn Financial Aid Settlement 2017–18. See https://devtracker.dfid.gov.uk/countries/PN/projects (accessed 18 October 2018).

European Commission (2006) *Halting the Loss of Biodiversity by Sustaining Service for Human Well-being*. (COM 216 final). Brussels: EU.

Foreign & Commonwealth Office (2012) *The Overseas Territories: Security, Success and Sustainability*. London: The Stationery Office.

Gardner, A. (2017) Brexit, boundaries and imperial identities: A comparative view. *Journal of Social Archaeology* 17 (1), 3–26.

Gilmartin, M., Burke, P. and O'Callaghan, C. (2018) *Borders, Mobility and Belonging in the Era of Brexit and Trump*. Bristol: Policy Press.

Graci, S. and Dodds, R. (2010) *Sustainable Tourism in Island Destinations*. London: Earthscan.

Graci, S. and Maher, P.T. (2018) Tourism. In G. Baldacchino (ed.) *The Routledge International Book of Island Studies* (pp. 247–260). Abingdon: Taylor & Francis.

Griffith, W. (2002) A tale of four CARICOM countries. *Journal of Economic Issues* 36 (1), 79–106.

Grydehoj, A. (2011) Making the most of smallness: Economic policy in microstates and sub-national island jurisdictions. *Space and Polity* 15 (3), 183–196.

Hall, S. (1990) Cultural identity and diaspora. In J. Rutherford (ed.) *Identity, Community, Culture, Difference* (pp. 222–237). London: Lawrence and Wishart.

Harmer, N., Gaskarth, J. and Gibb, R. (2015) Distant relations: Identity and materiality in elite discourse on Britain's Overseas Territories. *Global Society* 29 (4), 510–529.

Harrison, D. (2001) Tourism in small islands and microstates. *Tourism Recreation Research* 26 (3), 3–8.

Hendry, I. and Dickson, S. (2011) *British Overseas Territories Law*. Oxford and Portland, OR: Hart Publishing.

Hopkin, J. (2017) When Polanyi met Farage: Market fundamentalism, economic nationalism, and Britain's exit from the European Union. *The British Journal of Politics and International Relations* 19 (3), 465–478.

Jackson, B.G. (1996) Re-engineering the sense of self: The manager and the management guru. *Journal of Management Studies* 33 (5), 571–590.

Jordan, A., Wurzel, R.K. and Zito, A. (2005) The rise of the 'new' policy instruments in comparative perspective: Has governance eclipsed government? *Political Studies* 53 (3), 477–496.

Lewis, A. (2009) Pitcairn's tortured past: A legal history. In D. Oliver (ed.) *Justice, Legality, and the Rule of Law Lessons from the Pitcairn Prosecutions* (pp. 39–61). Oxford: Oxford University Press.
Lewis, P. (2016) The Repercussions of Brexit for CARICOM's cohesion. *The Round Table* 105 (5), 531–542.
MacLeod, D. (2013) Cultural realignment, islands and the influence of tourism. *Shima: The International Journal of Research into Island Cultures* 7 (2), 74–91.
McConnell, F. (2017) Liminal geopolitics: The subjectivity and spatiality of diplomacy at the margins. *Royal Geographical Society (with IBG)* 42, 139–152.
McConnell, F. and Dittmer, J. (2018) Liminality and the diplomacy of the British Overseas Territories: An assemblage approach. *Environment and Planning D: Society and Space* 36 (1), 139–158.
McElroy, J.L. and Pearce, K.B. (2006) The advantages of political affiliation: Dependent and independent small-island profiles. *Round Table* 95 (386), 529–539.
McElroy, J.L. and Hamma, P.E. (2010) SITEs revisited: Socioeconomic and demographic contours of small island tourist economies. *Asia Pacific Viewpoint* 51 (1), 36–46.
Megaw, N. (2017) Overseas territories warn of 'catastrophic' post-Brexit funding loss and Falklands risk. *The Financial Times*. See https://www.google.co.nz/url?sa=t&rct=j&q=&esrc=s&source=web&cd=&cad=rja&uact=8&ved=2ahUKEwjahbrqsaXqAhUd7XMBHaT_Cc4QFjAAegQIBhAB&url=https%3A%2F%2Fwww.ft.com%2Fcontent%2F6ce603d6-0ec4-3bc0-a7ca-0d893e1e69ac&usg=AOvVaw0vOvmDMsnQL_-OvYeiam-E (accessed 7 November 2018).
Pitcairn Island Tourism Department (2015) *Sustainable Tourism Development Master Plan 2015–2019*. Pitcairn Islands: Pitcairn Island Government.
Pitcairn Miscellany (2018) Pitcairn tourism and its South Pacific Island partners join forces at Seatrade Cruise Global 2018. *The Pitcairn Miscellany* 61 (3), 1–10.
Pratt, S. (2015) The economic impact of tourism in SIDS. *Annals of Tourism Research* 52, 148–160.
Putnam, R. (2000) *Bowling Alone: The Collapse and Revival of American Community*. New York: Simon & Schuster.
Scheyvens, R.A. and Momsen, J.H. (2008) Tourism and poverty reduction: Issues for small island states. *Tourism Geographies* 10 (1), 22–41.
Sharpley, R. (2012) Island tourism or tourism on islands? *Tourism Recreation Research* 37 (2), 167–172.
Stoll-Davey, C. (2017) *After Brexit: The Legal Relationship of the UK-OCTs with the EU*. London: University of London.
Trenwith, A. (2003) The empire strikes back: Human rights and the Pitcairn proceedings. *Journal of South Pacific Law* 7 (2), 1–7.
von Gesseneck, M.J. and Zieroth, G.H. (2017) Renewable energy feasibility study Pitcairn Island. See http://octa-innovation.eu/category/leslie-jaques/ (accessed 20 September 2018).
Weaver, D.B. (2017) Core–periphery relationships and the sustainability paradox of small island tourism. *Tourism Recreation Research* 42 (1), 11–21.
Wincott, D., Peterson, J. and Convery, A. (2017) Introduction: Studying Brexit's causes and consequences. *The British Journal of Politics and International Relations* 19 (3), 429–433.

12 Associate EU Citizenship: A Panacea for Loss of Fundamental Rights, Mobility and European Identity Post-Brexit?

Victoria Hooton

Introduction

Brexit has undeniably sparked a crisis of identity for many in the UK. It has already had a profound impact on the lives of British citizens and their families and relationships. Free movement is at the heart of the Brexit debate on both sides, with those who are pro-European advocating for the opportunities that free movement delivers to European Union (EU) citizens, and those who are more Eurosceptic demanding greater protection of national borders and territories. During these heated debates, the legal framework of EU free movement has been afforded little attention. EU citizenship and free movement is an area of EU and national law that is comprised of complicated and continually developing concepts. Citizenship is the crossroads of migration, residency, equal treatment and non-discrimination. These concepts contain complicated legal definitions, making their absence in Brexit debates understandable.

One way that citizenship has entered the Brexit stage is by the demand for an 'associate' EU citizenship, a mechanism whereby those who *feel* European and wish to *be* a part of the EU project, can retain their free movement and democratic voting rights within the EU. This mechanism *prima facie* appears to be a solid, functioning way of appeasing those of the UK electorate who wished to remain within the EU. This chapter will address the likelihood of associate citizenship ever truly living up to the emotional and social value attached to it. First, a review of the proposal for associate citizenship, and the narrative surrounding the concept is undertaken. Second, the framework for free movement at the EU level will be discussed to show that value judgements regarding citizenship are not always strictly accurate, and that, in reality, enjoyment of any associate citizenship rights would be limited. The final part of this chapter will assess how the social values and feelings of identity attached to the concept would compare to the reality of citizenship enjoyment for many EU nationals.

Introducing Associate EU Citizenship: Necessity, Narrative and Meaning

The demand for an 'associate citizenship' status is a direct result of the Brexit vote in 2016. The result of the referendum on the UK's continued EU membership, and the notification of withdrawal that followed, means the loss of EU citizenship for UK nationals. This has negatively impacted UK citizens who view their EU citizenship as a fundamental part of their identity, and those who rely on citizenship to secure their living and residency arrangements. Associate citizenship is seen as a way of mitigating this loss and insecurity.

Necessity

EU citizenship was introduced with the Maastricht Treaty in 1992. It is derivative, meaning it is governed by, ancillary to and very different from national citizenship. EU citizenship is conferred upon all individuals holding the nationality of a member state, as per Article 20 of the Treaty on the Functioning of the European Union ('TFEU' [2012] OJ C326/47). The member states control the scope of EU citizenship, by determining the rules for nationality of their territory. The rights provided by citizenship are primarily the ability to move and reside in another member state and the right to participate in EU elections, among other things.

Treaty on the European Union ('TEU' [2012] OJ C326/13) states that the treaties will cease to apply to a withdrawing member state as soon as a withdrawal agreement comes into action, or failing that after a period of two years (plus extensions) after notification of withdrawal. Because citizenship is bestowed upon UK citizens directly by the treaties, once those treaties no longer apply post-Brexit, UK nationals will no longer be EU citizens. Furthermore, as the treaties specifically tie EU citizenship to member state nationality, citizenship depends on 'member state' status. When a state withdraws from the Union, and is no longer a member state, those with the nationality of the ex-member state can no longer be considered 'citizens' of the EU (González, 2016: 797; Kochenov, 2016: 26; Van der Mei, 2018a).

The rights of many will be stripped away overnight. For the percentage of the UK who voted to stay within the Union, this could create immense distress. Around 4 million citizens face the greatest amount of uncertainty in their living arrangements, having taken part in the most fundamental aspect of EU citizenship: free movement of persons (Kostakopoulou, 2018: 866). Citizens will lose the protection of regulations, directives and judicial review from the Court of Justice of the European

Union (CJEU) that cements their residency rights in place, whether they are a UK citizen who loses 'citizenship' status, or an EU national residing in the (non-member state) UK.

In response to large numbers of current citizens being stripped of their rights against their will, certain members of the European Parliament advocated for the creation of a special status to be awarded to those individuals. Guy Verhofstadt, the representative of the European Parliament's views during Brexit talks, began to float the idea for an 'opt-in' citizenship in late 2016 (Stone, 2016). The idea is to create a new legal status, offering a package of rights specifically for ex-member state nationals, obtainable by paying a fee that would contribute to the EU budget (Goerens, 2017). This system would require a form of allegiance between an individual of an ex-member state and the EU, which would in return bestow upon those individuals the right to move and reside in other member states and the right to vote and stand for election in the European Parliament, within the member state in which they reside.

The European Parliament has not yet voted on a proposal to create an associate citizenship status, with the resolution being withdrawn before the vote (European Parliament Press, 2016). However, the proposal and its feasibility remain open to speculation. There is no open and shut case for associate citizenship, as rigorous academic debate exists as to what model of mobility is most suited to protect the rights of UK and EU nationals post-Brexit. There is a great deal of uncertainty regarding the form that associate citizenship should take (Fowles, 2018). For instance, whether it should be given to every UK national, or only those who have already exercised free movement (Kostakopoulou, 2018; Van den Brink & Kochenov, 2018; Van der Mei, 2018b). Moreover, whether the creation of a new status would require a treaty change at the EU level is debatable (Piris, 2016; Roeben *et al.*, 2018; Van den Brink & Kochenov, 2018). The creation of a special status may be seen to fly in the face of the UK's democratic decision to withdraw from the EU (Kochenov, 2018; Van den Brink & Kochenov, 2018: 14; Van der Mei, 2018b). This is particularly true because the Brexit campaign has become so entrenched in anti-immigration and anti-free movement rhetoric. Furthermore, there is a lack of political will to make associate citizenship happen, it is against the EU's interest as it makes withdrawal more attractive (Van den Brink & Kochenov, 2018: 16–19) and national governments may see it as a loss of control over the scope and nature of citizenship (Fowles, 2018).

Narrative

Two members of the European Parliament (MEPs) are advocating for associate citizenship: Guy Verhofstadt (Belgium) and Charles Goerens (Luxemburg). The two view associate citizenship as a way of rectifying Brexit-related grievances for British citizens, who ultimately get no say on whether they retain their EU citizenship. Goerens (2017) notes that some British citizens are concerned by Brexit, and their loss of citizenship is 'a consequence they are not prepared to accept'. Verhofstadt (2018) remarked that he 'will continue to push for recognition that millions of UK citizens are having their European rights taken away from them against their will. Europe should recognise this, in [his] opinion'. Both MEPs mention the overwhelming communication they receive from UK citizens, expressing their angst over losing their European identity without any recourse (Goerens, 2017; Verhofstadt, 2017).

Academics (Kostakopoulou, 2018; Roeben *et al.*, 2017; van den Brink & Kochenov, 2018) have reflected on the fundamental status[1] that has been awarded to EU citizenship by the CJEU and noted this as a reason for its continued significance post-Brexit. The arguments differ regarding what should be done to protect this fundamental status, but provide a consistent narrative of the fundamentality of citizenship. A team of academics at Swansea University (Roeben *et al.*, 2017) concluded that citizenship of the EU is a fundamental status of individuals, and therefore cannot be removed even in the event of Brexit, under EU law or international law. The status quo would be protected, meaning that those who have exercised free movement would have some legitimate expectation to be able to do so in the future, but would have less protection under EU law (Roeben *et al.*, 2017). The authors see 'associate citizenship' as an alternative model to the simple 'continuity' of citizenship, but still argue that *if* citizenship is stripped from UK nationals post-Brexit, the EU has a positive obligation to create associate citizenship, to protect the rights of its subjects from interference by third parties. The authors also entrench this thinking in human rights principles, noting that certain citizenship rights are guaranteed in the EU's Charter of Fundamental Rights, and referencing the international duties to respect, protect and fulfil human rights (Roeben *et al.*, 2017: 28).

Dawson and Augenstein (2016) have suggested 'de-coupling' EU citizenship from national citizenship, to truly recognise the fundamental status of EU citizenship, protecting that status from the political preferences of governments. This would make it so that member states would retain control over granting EU citizenship, but the decision to withdraw it would be down to the individual citizen. Any decision to withdraw as a member state would not affect the EU citizenship of its nationals. The authors ground their argument on CJEU cases that have suggested fragmentation of the link between member state nationality and EU citizenship. Citizenship's independent and fundamental legal status lends credence to its independence from nationality. The authors also offer a broader view that this *should* be the case, because the EU project should be more centred on the needs and rights of citizens.

Kostakopoulou's (2018) suggestion of a 'protected status' for citizens who have exercised free movement (to and from the UK) encapsulates elements of practicality and fundamentality. The author recognises the need to protect the status quo for those most affected by Brexit, rather than all UK citizens. However, she underpins the rationale for freezing current free movement rights, with the 'fundamental status' of citizenship, as interpreted by the CJEU. Although Kostakopoulou laces her argument with fundamentality to strengthen its legal basis, the argument is ultimately in favour of a 'safety net' for those affected by Brexit in a practical rather than an emotional sense. The author does not maintain the suggestion that citizenship is so fundamental that it is inviolable, but she does make clear that she disagrees with any suggestion that Brexit highlights the fragility of EU citizenship.

Some academics have relied upon human rights law, especially the European Convention on Human Rights (ECHR, Article 8), to show that citizenship rights are a fundamental aspect of social identity (Alegre, 2016) and that residency rights in particular are tied to the human right to respect for family life and dignity (González, 2016), bringing the loss of EU citizenship possibly under the remit of the ECHR. These narratives also highlight the position of citizenship as something greater than a legal practicality, making up part of a person's social identity. Narrative from UK citizens does show a certain level of distress regarding the loss of citizenship and access to free movement. This is evidenced by the use of democratic and legal tools for change, as well as responses to questions during interviews undertaken by research centres.[2] UK nationals have shown direct defiance against their loss of EU citizenship by using an online petition to pressure the British and EU parliaments to make associate EU citizenship available (Change.org, 2016).

In the Netherlands, a group of UK nationals have even taken up legal proceedings (de Rechtspraak, 2018) to ensure that the Netherlands recognises its positive obligation to respect the citizenship rights of Britons and their families residing there. The distress that is experienced by these claimants relates to the practical changes of EU citizens' everyday lives, but the legal arguments made were for protection of their fundamental rights from ongoing and further detriment. The Amsterdam court also recognised this fundamentality (de Rechtspraak, 2018).

Furthermore, the EU citizens' initiative tool has been used to pressure the EU Commission (2018) into protecting EU citizenship and rights. The Commission registered the initiative, and stated that EU law regulates the free movement of third country (non-member state) nationals, which can confer 'similar rights' to EU citizens. However, the UK nationals in this case wish to retain 'EU citizenship *and* rights'. There is a level of divorce between the status of citizenship and the rights that it gives. Stakeholder interviews with UK nationals also highlight that UK citizens find their EU citizenship as 'an inseparable part of their personal identity, regardless of where they currently live' (Austin-Greenall & Lipinska, 2017: 19).

In terms of what is meant by this European *identity*, a study by Collins and O'Reilly (2018) found that British citizens that had exercised their right to free movement embraced the concept of EU citizenship in a symbolic way. They tied their EU citizenship to values of openness and appreciation for cultural diversity that is closely related to the individual's sense of belonging and identity. The loss of citizenship rights sparks emotional distress that does not just stem from practical concerns around residency status, but also from a loss of opportunity, social mobility and social goods such as cultural pluralism and new ideas and experiences (Collins & O'Reilly, 2018: 7). Panellists for the study gave the impression that they were concerned with their future plans as well as their current residency status, and that the loss of citizenship and free movement would also be a loss of 'a sense of who they are, and who they can be' (Collins & O'Reilly, 2018: 8).

What can be drawn from all three of the narratives considered is that citizenship is viewed as something more than the sum of its parts, which are generally legal provisions formed in different treaties, regulations and directives. It forms a key part of the identity of its beneficiaries that is somewhat divorced from the actual legal rights that it provides to them. It is also clear that 'there is a great lack of comprehension amongst UK citizens, as to what specific rights and benefits derive from EU membership' (Austin-Greenall & Lipinska, 2017: 20).

From the preceding discussion, it appears that legal protection of certain rights via a treaty or other agreement is not enough to reduce the distress felt by some UK citizens. Something more robust is desired, which would be 'associate citizenship'. However, associate citizenship would be a shallow answer to what is effectively a much deeper question of identity, belonging and social value. The social goods recognised in the narratives on loss of identity are not always direct results of EU citizenship. The realisation of citizenship rights can be complex, exclusive and can often lead to unfair and arbitrary results. It is not the purpose of this chapter to dispute the recognition of a European identity, but instead to highlight the different experiences of EU citizenship and the fragmentation between the broader feelings of loss experienced by UK nationals stripped of their citizenship, and the actual makeup of the free movement framework that associate citizenship would bridge access to.

Meaning

Van der Mei (2018b) states that it is uncertain what advocates of associate EU citizenship want. Is the demand that there should be a simple freezing of the status quo, or that future rights should be solidified? The narratives explored in this chapter evidence that certain UK citizens want to retain citizenship rights and their *status* as a citizen, i.e. their identity.

The narrative from Goerens and Verhofstadt makes it clear that associate citizenship would be for *all* those who want to be involved in the EU project. As this chapter is taking a socio-legal approach to assessing the proposal for associate citizenship, it will now leave behind the legal and political complexities of the differing models proposed, including their legal tenability and political feasibility. Instead, this chapter will take the political proposal from the European Parliament – the full 'associate citizenship' bundle of rights for UK nationals regardless of free movement status – which is the model that should appeal to nationals aggrieved by their loss of European identity. The purpose of the next section is to show that greater value than is necessarily warranted in terms of equality, social mobility and inclusion is placed upon the idea of EU citizenship.

Free Movement and the True Value of Citizenship: A Class Enterprise

Free movement is regulated by EU primary and secondary legislation. The rights to move and to reside in another member state are enshrined in the treaties, the highest form of EU law. The substance and limitations of those rights are fleshed out in directives and regulations. It is important to note that because associate citizenship would be an entirely new status, it could involve more limitations than the current free movement rights afforded by EU citizenship. However, as the concept 'associate citizenship' has been floated because of the distress of losing currently active rights, this chapter considers that the free movement framework would apply the same to associate citizens as to EU citizens. Furthermore, Roeben *et al.* (2018: 37) find that the current EU free movement framework will be the baseline for any future agreement or Union legislation.

Regardless of the feasibility or likelihood of 'associate citizenship' mirroring the free movement legal framework as it currently stands, the question addressed in this chapter is whether the fundamental loss argument, and the emotional value attached to citizenship is justifiable, and whether a mirroring of those rights would truly rectify the loss felt. It is possible that the identity loss, and the fundamentalism, is attached to the *status* of citizenship and not the rights it conveys via the free movement framework, but it seems unjustifiable to separate the two. As Spaventa (2017) notes, citizenship should be understood through its scope, as the legal developments regarding citizenship will adjust how it is viewed.

The free movement legal framework, including relevant case law developments, highlights the boundaries of the fundamental status of citizenship and the opportunities it offers. As per Article 21(2) of the TFEU, all EU citizens are free to move and reside in any member state. The CJEU has itself stated that, 'Union Citizenship is destined to be the fundamental status of the nationals of the Member States, enabling those who find themselves in the same situation to enjoy the same treatment in law irrespective of their nationality' (Case C-184/99 *Grzelczyk*, 2001: 31). This contributes to the narrative of fundamentality surrounding citizenship, and certainly looks like the traditional post-French Revolution notion of citizenship: that all persons should be equal before the law (Kochenov, 2017).

However, the TFEU also notes that citizenship rights are subject to limitations and conditions. Interpretations of these limitations and conditions in EU law throw into question the fundamentality of EU citizenship, and therefore its related identity and culture. The main source of rights and conditions are contained in the Citizens' Rights Directive (2004/38/EC; henceforth 'CRD'). The CRD makes it clear that not all citizens are equal, and the CJEU cements this in case law. The rights and duties attached to citizenship differ according to the length of time spent in another member state and the personal situation of the citizen.

Up to three months residency

The CRD requires all member states to allow EU citizens to enter their territory with very little formality in the first instance. This makes travel throughout the Union relatively easy for tourism, reducing the administrative burdens and costs for citizens wishing to briefly visit another member state. For the first three months of residency in a member state, there are also very few formalities. There are no conditions or duties to be met by the citizen, so long as they have a valid identity card or passport. Conversely, there are no substantial rights beyond the right to stay within this period.

Three months to five years

The period between three months and five years imposes the highest duties and conditions on citizens resident in another member state. It exemplifies the inequality experienced by different types of citizens. During this time, the CRD differentiates the conditions applicable to an individual's stay depending on whether they are a worker or self-employed, a jobseeker, a student or self-sufficient.

Workers and self-employed individuals have the greatest right to stay in a member state; they have a treaty right to move and reside elsewhere, with the same social and tax advantages as nationals (Art. 45 TFEU, Regulation 492/2011). The positioning of the workers as apex citizens hints at a non-egalitarian form of citizenship. Furthermore, O'Brien (2016: 939,953) finds that because there are no 'bright lines' demonstrating the difference between economic activity and inactivity, member states have the freedom to distort the meaning of 'work', and can choose to ignore poorly paid work, declining equal treatment to those on zero-hours, flexible working and gig economy contracts. O'Brien (2016: 956) notes that EU nationals in the UK are most likely to be concentrated in low-paid work, and that many are working 20 hours a week but are considered not to have 'worker' status because of poor pay.

Jobseekers are required to be looking for work with a genuine chance of engagement to retain lawful residency. This can essentially oblige a citizen to have a job offer with a substantial enough wage attached (O'Brien, 2016). Furthermore, there is no social support for jobseekers during this search (Art. 24(2) CRD). Those who have previously worked have protected rights as workers, which vary depending on how long they have been employed for (Art. 7(3) CRD). If a citizen has worked for less than a year, then they lose any protection that worker status affords them after six months. This is regardless of the degree of connection with the member state where they live. In *Alimanovic* (Case C-67/14, 2017), an ex-worker lost her rights to social assistance (while still legally resident in Germany). O'Brien (2016) highlights Ms Alimanovic's strong economic links to Germany, and Kramer (2016) suggests her personal history as a refugee in Germany, fleeing violence. The *Alimanovic* case illustrates how citizenship rights can appear exclusive, arbitrary and unfair for those without stable and substantial employment.

Students are required to have comprehensive sickness insurance, be enrolled and studying at an educational institution in the member state and have sufficient resources so they will not become a burden on the social assistance system. Already there is a clear indication that *wealthy* enough students are welcome to use free movement. Article 24(2) CRD also dictates that students are not entitled to maintenance grants or loans unless they are permanently resident in a member state (see below), or if they are a worker or a family member of a worker. The CJEU has fiercely upheld this rule in *Förster* (Case C-158/07, 2008), where a citizen had forged substantial social and economic links with her member state of residency but had no entitlement to maintenance grants. This decision has been subject to much academic criticism (Barnard, 2005; Golynker, 2009; O'Leary, 2009; Van der Mei, 2009). The effect of this system is that it is mainly middle-class, educated students who can make use of free movement for the purposes of study (Barnard, 2005: 1482). It is unsurprising that, in the body of the *Förster* judgment (Case C-158/07, 2008), the court did not reiterate its statement that citizenship is to be the 'fundamental status' of EU nationals.

All other economically inactive individuals need to have 'sufficient resources' according to Article 7(1)(b) CRD. This requirement again demonstrates the commitment to enabling only wealth mobility within the Union. This disparity in access to movement is further exacerbated by the term 'sufficient resources' being relatively undefined in the CRD. In a landmark decision, *Dano* (Case C-333/13, 2016), the CJEU appeared to permit an interpretation of 'sufficient resources' meaning a citizen does not attempt to claim social benefits. The result of this is that citizens who do require financial assistance will put their entire right to residency at risk by attempting to claim it, because it will throw into question their 'sufficient resources'. The case attracted much media coverage, as it was seen as a victory against abuse of free movement rules and 'benefit tourism'.

These developments have occurred over the course of the last decade. There was a time when the CJEU was moulding citizenship into something substantial, noting its fundamental status and requiring member states to recognise a degree of solidarity with EU citizens (O'Brien, 2016: 940). Spaventa (2017: 206) observes that this is a 'reactionary phase' for the CJEU, undermining the 'fundamental' nature of EU citizenship. EU citizenship is currently shaped by nation-centric politics and mistrust of free movement (Nic Shuibhne, 2015; O'Brien, 2016).

Permanent residency

After a period of five years of legal residency (i.e. conforming with the criteria discussed above), citizens obtain the right to permanent residency, and the conditions no longer apply. There are some limits on this, but the important aspect to note for this chapter is that the strongest rights to residency are based upon a requisite (and arguably rather long) length of time spent in a member state.

It is important to note that this is a diluted overview of citizenship and free movement, which is inherently complex and involves more caveats and exceptions than there is scope for in this chapter. What the above does highlight is that there are significant inequalities in the way citizenship is enjoyed by those who hold the *fundamental* status. Free movement is a right for the sufficiently wealthy, many of whom will enjoy mobility regardless of the free movement framework (albeit with higher costs and administrative burdens).

Associate Citizenship: A Panacea for Loss of Rights and Identity?

The previous section evidences a direct divide between the citizenship opportunities enjoyed by those who can take part in the market economy and those who cannot. Any hypothetical 'associate citizenship' would be beneficial only for the people with the resources to access it. Furthermore, the reactionary case law discussed above shows how citizenship is not strong or fundamental enough to outweigh the sway of national politics. The narrative that EU citizenship is so fundamental that it should not be affected by any governmental political preference has been effectively undermined by the legally defined boundaries of EU citizenship itself.

Associate citizenship rights would ensure the continuation of short-term travel stays, which may be looked on favourably by a large portion of the UK electorate. It would also secure the rights of those who have found stable and substantial employment, and those who are wealthy enough to relocate. However, it should be remembered that a treaty between the UK and EU would achieve these things (Van den Brink & Kochenov, 2018). It is the status of citizenship that is paramount to the requirement for 'associate' citizenship, to retain the European identity of openness, social mobility and cultural diversity. It seems likely, however, that retention of access to the free movement framework via associate citizenship would hinder, rather than encourage, those values.

Associate EU citizenship would not harness the respect for cultural diversity and pluralism that some UK nationals have (quite understandably) attached as a value to free movement. A new status for UK nationals could be considered to be British exceptionalism, setting UK citizens apart from other third-country nationals. Third-country nationals already resident within the EU would have much deeper connections to the territory than UK nationals who remain in the UK, but are nonetheless offered associate citizenship. The citizenship framework does not support the perceived value of openness, freedom and experience that has also been expressed as forming part of certain EU citizens' identities, as it connects the strength of rights to the length of time spent in a member state. This encourages long-term resettling, more than mid-term experiencing of work and life in other countries, to the detriment of citizens who are more highly mobile.

By far the greatest issue with associate citizenship is its effect on socioeconomic divides. Citizenship does not automatically create greater social mobility. As Kochenov (2018) notes, equality before the law for EU citizens is extended to 'the "good citizen", who is economically productive, mobile, and successful', a sentiment which is shared by Spaventa (2017). Only those individuals who can comply with the conditions of the CRD may enjoy citizenship. So great is the tie between wealth and citizenship that Spaventa (2017: 223) states she would rather have *no* citizenship than one based on this level of exclusion and discrimination. Since associate citizenship would come at a price, it would cement and intensify the connection between socioeconomic status and enjoyment of free movement. It would also increase social tension in the UK (Fowles, 2018), perhaps further disenfranchising large parts of the populous. Shaw (2018) notes that 'part of the reason for the Brexit vote was precisely that EU citizenship was not recognised as a social fact by the majority of voters', which makes it difficult to conclude that EU citizenship makes up a fundamental part of the identity of those who hold that status.

A caveat is necessary as it is clear from the narratives above that citizens have gained opportunities from the free movement framework that they may not have otherwise had. This chapter does not suggest that EU citizenship never increases social mobility, merely that it does not *always* do so, and cannot guarantee a consistent level of enjoyment of rights. Therefore, it cannot truly make up part of a nation's fundamental or social identity.

Conclusion

This chapter has established the existence of a narrative that the loss of mobility rights for UK nationals post-Brexit fundamentally undermines their social and cultural identity relating to EU citizenship. The answer to this, as suggested by Verhofstadt and Goerens, would be to create an associate citizenship so that British nationals can retain their fundamental right to take part in the EU project. The current state of EU law throws the fundamental nature of citizenship into question and highlights underlying issues with the free movement framework that could be exacerbated by the existence of associate citizenship. The value attached to that status and the identity that it creates are not reflective of the nature of legal reality, which is increasingly enjoyed by those with a particular socioeconomic background.

The broader context of this is that there is a general misunderstanding of what EU citizenship and mobility is, what it does and what it should be. This misunderstanding permeates current dialogue on the status of the Union. It underpins the entire edifice of the referendum, Brexit, feelings of loss and the shifting nature of EU citizenship. Difficult questions about the fundamentality of rights arise because of a transnational level of miscommunication between citizens, executives and legislatures. There is a clear lack of dialogue between large proportions of Union citizens and the institutions of the Union itself.

It is no surprise that there is fragmentation between perceptions of European identity and the reality of EU citizenship as it is accessed and experienced. There is little to no roadmap for citizenship (Spaventa, 2017), in terms of what its aim and ultimate destination are. This leads individuals to attach their own value and meaning to it, which builds divergent views of citizenship and European identity. Those views will be positive for the wealthy and mobile, and negative for those who are not. Overall, it is unlikely that either view of European identity and citizenship represents the true scope of its legal status.

There is nothing stopping EU citizenship progressing into something more accessible by all in the future. Kostakopoulou (2018) notes that there are institutional changes that could more fully realise the potential of EU citizenship, and Kochenov (2017: 227) notes that if citizenship were less about free movement, it could be embedded into more egalitarian issues like democracy and fundamental rights. It is certainly necessary to make free movement, or citizenship, fairer to truly start to build a common sense of identity among the peoples of the EU. However, it remains important that we recognise the limitations of citizenship as it is currently experienced to better evaluate the proposals for associate citizenship.

Notes

(1) Fundamentalism, in the sphere of EU law considered here, indicates rights and legal protections that are of greater importance than others. The 'fundamental' right to

free movement, for instance, cannot lightly be impinged upon, and restrictions on this right are to be construed as narrowly as possible, to avoid undue interference.
(2) Discussed below, studies by Austin-Greenall and Lipinska (2017) and Collins and O'Reilly (2018) involved interviewing UK citizens to determine their concerns and thoughts about their EU citizenship and freedom of movement rights.

References

Alegre, S. (2016) Written evidence for House of Lords EU Justice Sub-Committee – Brexit Acquired Rights Inquiry (AQR0007) HL Paper 82.

Austin-Greenall, G. and Lipinska, S. (2017) *Brexit and Loss of EU Citizenship: Cases, Options, Perceptions*. Brussels: ECAS.

Barnard, C. (2005) Case Comment Case C-209/03, R (On the Application of Danny Bidar) v. London Borough of Ealing, Secretary of State for Education and Skills, Judgment of the Court (Grand Chamber). *Common Market Law Review* 42, 1465.

Case C-67/14 *Alimanovic* (2015) EU:C:2015:597.

Case C-333/13 *Dano* (2014) EU:C:2014:2358.

Case C-158/07 *Förster* (2008) ECR I-08507.

Case C-184/99 *Grzelczyk* (2001) ECR I-06193.

Change.org (2016) See https://www.change.org/p/do-you-want-personal-eu-citizenship-send-a-message-to-the-european-parliament (accessed 27 August 2019).

Collins, K. and O'Reilly, K. (2018) *What Does Freedom of Movement Mean to British Citizens Living in the EU-27? Freedom, Mobility and the Experience of Loss*. London: Goldsmiths.

Consolidated version of the Treaty on European Union [2012] OJ C326/13.

Consolidated version of the Treaty on the Functioning of the European Union [2012] OJ C326/47.

Dawson, M. and Augenstein, D. (2016) After Brexit: Time for a further decoupling of European and national citizenship? VerfBlog, blog post, 14 July. See https://verfassungsblog.de/brexit-decoupling-european-national-citizenship/ (accessed 29 August 2019).

Directive 2004/38/EC on the right of citizens of the Union and their family members to move and reside freely within the territory of the Member States [2004] OJ L158/77.

EU Commission (2018) C(2018) 4557 final Commission Decision on the proposed citizens' initiative entitled 'Permanent European Union Citizenship', Brussels, 18 July.

European Parliament Press (2016) See http://www.europarl.europa.eu/unitedkingdom/en/media/euromyths/associatesitizenship.html (accessed 29 August 2019).

Fowles, S. (2018) Associate EU citizenship: A brief assessment, Foreign Policy Centre, blog. See https://fpc.org.uk/associate-eu-citizenship-a-brief-assessment/#_ftn3 (accessed 19 August 2019).

Goerens, C. (2017) Pan European, April. See https://thepaneuropean.eu/2017/04/30/associate-eu-citizenship-interview-charles-goerens/ (accessed 29 August 2019).

Golynker, O. (2009) Case Comment C-158/07 Förster. *Common Market Law Review* 46, 2021.

González, G.M. (2016) BREXIT Consequences for Citizenship of the Union and Residence Rights. 23 MJ 5.

Kochenov, D. (2016) EU Citizenship and Withdrawals from the Union: How Inevitable Is the Radical Downgrading of Rights? University of Groningen Faculty of Law Research Paper Series No. 23/2016.

Kochenov, D. (2017) European Citizenship and Its new Union: Time to Move on? In D. Kochev (ed.) *EU Citizenship and Federalism: The Role of Rights* (pp. 226–242). Cambridge: Cambridge University Press.

Kochenov, D. (2018) Misguided 'associate EU citizenship' talk as a denial of EU values, VerfBlog, blog post. 1 March. See https://verfassungsblog.de/misguided-associate-eu-citizenship-talk-as-a-denial-of-eu-values/ (accessed on 27 August 2019).

Kostakopoulou, D. (2018) *Scala civium*: Citizenship templates post-Brexit and the European Union's duty to protect EU citizens. *Journal of Common Market Studies* 56 (4), 854–869.
Kramer, D. (2016) Earning social citizenship in the European Union: Free movement and access to social assistance benefits reconstructed. *Cambridge Yearbook of European Legal Studies* 18, 270–301.
Nic Shuibhne, N. (2015) Limits rising, duties ascending: The changing legal shape of Union citizenship. *Common Market Law Review* 52, 889–937.
O'Brien, C. (2016) *Civis Capitalist Sum*: Class as the new guiding principle of EU free movement rights. *Common Market Law Review* 53, 937.
O'Leary, S. (2009) Equal treatment and EU citizens: A new chapter on cross-border educational mobility and access to student financial assistance. *European Law Review* 34 (3), 612–627.
Piris, J-C. (2016) EU citizenship deal for British nationals has no chance, say experts. *The Guardian*. See https://www.theguardian.com/uk-news/2016/dec/12/eu-citizenship-deal-for-british-nationals-has-no-chance-say-experts (accessed 27 August 2019).
Rechtspraak de (2018) Court of Amsterdam, 200.235.073/01. See https://uitspraken.rechtspraak.nl/inziendocument?id=ECLI:NL:RBAMS: 2018:605 (accessed 28 August 2019).
Regulation (EU) No. 492/2011 of the European Parliament and of the Council of 5 April 2011 on freedom of movement for workers within the Union. [2011] OJ L141/1.
Roeben, V., Snell, J., Minnerop, P., Telles, P. and Bush, K. (2017) *The Feasibility of Associate EU Citizenship for UK Citizens Post-Brexit*. Swansea: Swansea University.
Shaw, J. (2018) EU citizenship: Still a fundamental status? RSCAS 2018/14, EUI Italy.
Spaventa, E. (2017) Earned citizenship: Understanding Union citizenship through its scope. In D. Kochev (ed.) *EU Citizenship and Federalism, the Role of Rights* (pp. 204–225). Cambridge: Cambridge University Press.
Stone, J. (2016) EU negotiators will offer Brits an independent opt-in to remain EU citizens, chief negotiator confirms. *The Independent*, December. See https://www.independent.co.uk/news/uk/politics/brexit-eu-citizenship-keep-freedom-of-movement-guy-verhofstadt-chief-negotiator-opt-in-passports-a7465271.html (accessed 17 September 2019).
Van den Brink, M. and Kochenov, D. (2018) A Critical Perspective on Associate EU Citizenship after Brexit. DCU Brexit Institute – Working paper N. 5 – 2018, University of Groningen Faculty of Law Research Paper No. 12/2018.
Van der Mei, A.P. (2009) Union citizenship and the legality of durational residence requirements for entitlement to student financial aid. *Maastricht Journal of European & Comparative Law* 16, 477–496.
Van der Mei, A.P. (2018a) Maastricht University blog, 22 October. See https://www.maastrichtuniversity.nl/blog/2018/10/brexit-and-citizenship-i-retention-eu-citizenship (accessed 26 August 2019).
Van der Mei, A.P. (2018b) Maastricht University blog, 23 October. See https://www.maastrichtuniversity.nl/blog/2018/10/brexit-and-citizenship-ii-associate-eu-citizenship (accessed 26 August 2019).
Verhofstadt, G. (2017) The EU will defend its interests in the Brexit talks, but will also be generous to British citizens. *The Independent*. See https://www.independent.co.uk/voices/the-eu-will-defend-its-interests-in-the-brexit-talks-but-will-also-be-generous-to-british-citizens-a7674371.html (accessed 29 August 2019).
Verhofstadt, G. (2018) The Pan European, February. See https://thepaneuropean.eu/2018/02/11/interview-guy-verhofstadt-citizens-rights/ (accessed 29 August 2019).

13 Brexit and Post-Globalisation Era: Walking into Unknown Geography

Reza Masoudi

Introduction

In the company of my six-year-old son, I was walking into a polling station for the referendum on the UK's membership of the European Union (EU) in Wanstead, London, in June 2016. While my son was asking about the referendum, I experienced a feeling of déjà vu. I recalled my 10-year-old self walking into a polling station in the spring of 1979 in Iran when the Islamic Republic referendum was being held. In both cases, people cast their opinion about new political makeups that were little more than a title. Iranian voters were supposed to say 'yes' or 'no' to establishing an Islamic Republic, with a constitution that was yet to be written. In summer 2016, while everybody had an assumption about the meaning of Brexit, nobody knew its actual meaning, not even those who were against leaving the EU. Theresa May's meaningless famous phrase: 'Brexit means Brexit', well reflects the ambiguity of Brexit. Since then, there has been no progress on establishing the meaning of Brexit. By the end of March 2019, when the UK was supposed to have left the EU, the then Prime Minster Theresa May extended Article 50 to delay Brexit, and Parliament was in a historic crisis based on its failure to reach an agreement over delivering Brexit. After more than three years of public debates, political turbulence and moving Boris Johnson into Number 10 Downing Street as prime minister, promising to deliver Brexit by the end of October 2019, still nobody knows what Brexit holds for the UK. The narrow majority result of the referendum (51.89% vs 48.11%) certainly polarised the UK at an historic level, and the political rift is going to keep widening.

The UK's current historic political crisis is not merely about the dispute over how to deliver Brexit. It also reflects the *anxiety* and *excitement* of walking into a new geographical order. The outcome of the referendum signifies the end of globalisation, as we know it. We are clearly in a major transitional period, moving towards a new era when the global geography, the spatial pattern of markets, the idea of the nation-state and the notion of sovereignty will be changed. Surely history would not be reversed, and the landscape of the global order would not look like the pre-globalisation period or colonial era, those times that some people may romanticise. Nonetheless, the perception of the colonial world clearly played an active role during the referendum campaign to envisage the past and the future of the UK's global position. This perception is neither factually accurate nor historically comprehensive. However, such a perception is not supposed to be accurate; it is about how a sociopolitical group has decided to narrate and idealise a glorious past to envisage a future.

The rhetoric during the referendum campaigns was full of terminologies with geographical connotations such as globalisation, trade, sovereignty and immigration. Nonetheless, there has been very little awareness about the sheer impact of Brexit on the complex geography in which we are living. While Brexit will transfigure the geography of the UK, there was a clear ignorance about the impact of Brexit on the UK's political geography. Who thought that the Irish border would become the most critical issue in Brexit negotiations? Any other exit from the EU could be delivered much easier than Brexit. But the complex political geography of the UK makes delivering Brexit such a complicated task.

The geography that we are living and acting on does not have only one layer, we are living in multiple geographies that are interwoven. If you live in Northern Ireland, your life is unfolded not only in the geography of Northern Ireland, but also in the island of Ireland, the UK, the EU, the Commonwealth and indeed in the geography of the globalised world. These layers of our geography – or multiple geographies – have been built up, folded and twisted throughout the globalisation era, the Cold War, the postcolonial era, the two world wars, the colonial era and beyond, shaping a complex spatial configuration. What makes delivering Brexit difficult is not just the complexities of getting out of the EU, but rather the need to reconfigure the UK's geography of which EU membership is only one of its interwoven layers.

In his speech, delivered at the Geographical Society in 1912, George Curzon of Kedleston (1915: 155–156; cited by Said, 1979: 2015) explained: 'By the aid of geography, and in no other way, we do understand the action of great natural forces, the distribution of population, the growth of commerce, the expansion of frontiers, the development of states, the splendid achievement of human energy in its various manifestations'. Geography does not merely refer to the location of things and places; it is about the distribution of human activities on physical geography. It is about our technological ability and ideological intentions in spatial arrangements of power, divisions of labour, production and consumption, all of which are subject to changes throughout history. Therefore, understanding the genealogy of our geography is crucial to comprehend its dynamics and evolution.

This chapter firstly articulates the landscape in which the EU referendum and the demand for Brexit unfolded, a landscape that globalisation and neoliberalism[1] had a great role in its configuration. Then, the discussions review the genealogy of globalisation and its geographical dynamics. I shall explain that while globalisation and neoliberalism are two layers of our geography, our lived experiences do not necessarily distinguish the two since they are entwined. We seamlessly live and act on multiple layers of geography, often not aware of each individual layer. The seamless lived experience of multilayered geography partly explains why populist politicians easily created an illusion that the working-class community of, for example, Sunderland and the capitalist elites who backed the Leave campaign have a common interest. I will explain that the layers of such geography are not configured as a three-dimensional system, but rather are constituted as a topological space, in which layers are folded and twisted; therefore, we have a seamless lived experience across multiple layers.

Globalisation: An Elephant in the Room?

The process of globalisation facilitated the West in conquering the world by liberal democracy as Fukuyama (1989) described it, or assisted Western capitalism moving out of its periodical crisis as David Harvey (e.g. 2006) argued. The globalisation process has maintained and expanded the domination of the West across the globe in the postcolonial era. The end of the Cold War abolished the borders between the West–East camps and facilitated neoliberal ideology to conquer the world economy by a capitalist free-market. While some scholars have been busy with ideologically charged analyses, other scholars have been warning about the sociopolitical consequences of the globalisation process, entangled with neoliberal ideology. Fukuyama celebrated the era of ultimate Western hegemony, whereas some others pronounced globalisation as neocolonialism, a form of colonialism that offers the West the power and influence without the responsibility (e.g. Young, 2001: 44–56). However, in *Losing Control*, Saskia Sassen (1996) warned that the globalisation process undermined two distinctive features of modern states: sovereignty and exclusive territory.

Neoliberalism was the ideology of globalisation, under which the nation-states gave up their sovereignty in privilege of international corporations that do not function in the interest of nations, but rather for capitalist elites. The Western hegemony of the colonial era was reproduced by Western corporations during the postcolonial era. In short, colonial imperialism was privatised. However, the rise of China is gradually diminishing Western dominance in the global order. The gravity of

China has gradually shifted the centre of global geography towards the East. China used to be one of the main sources of cheap labour and the production house of Western corporations. Now, it is clearly threatening the position of the West in the global landscape of advanced technologies. Although we have experienced a shift, not a change, this is very alarming for capitalist elites and right-wing politicians, who believe that Western dominance is in the interests of humanity.

The globalisation process saved Western capitalism during the second half of the 20th century by producing a geography in which markets and production processes were interconnected across national borders. This new global order brought an economic boom under neoliberalism, but then the financial crash of 2008 occurred, another unavoidable periodical crisis for capitalism as Harvey (e.g. 2006) would describe it. This financial crisis quickly spread like wildfire across national borders since deregulated markets around the world are interconnected, while nation-states had very limited abilities to control the crisis. The crisis was so serious that some even rushed to announce it as a sign of the end of capitalism. This was obviously not the case. The outcome was that the UK government bailed-out British banks and introduced a harsh austerity policy that put huge pressure on public services and the quality of life of the middle and working classes, leading to public frustration and dissatisfaction. This situation is well described by the classical political phrase: *Socialism for the Capitalist, Capitalism for the poor*.[2]

The transnational corporations not only reconfigured the geography of the market and production, but also created a new global pattern of immigration to redistribute human resources according to their needs. Sassen (2001) has argued that there is a duality in the geography of globalisation: It is simultaneously dispersed and extremely centralised. Transnational corporations disperse the geography of production to reach low-paid labourers in the Global South. On the other hand, their headquarters are located in a few cities known as global cities, where they need to allocate highly skilled and specialised individuals. Therefore, international corporations demand an immigration policy that eases the redistribution of highly specialised individuals across national borders. Not only transnational corporations but also international bodies, such as the United Nations (UN) and the EU, have contributed to undermining governments' powers to control their borders by implementing refugee and asylum conventions. Moreover, the process of globalisation *per se*, also contributes to the plea for global immigration due to uneven global development.

By the early 21st century, the scale and pattern of immigration created a new social landscape in major Western cities including London. Steve Vertovec (2007) has described the new social landscape of London as a 'super-diverse' status, since socioethnic diversity super-exceeded all previous experience. He argued that while immigrants were predominantly from a few countries, e.g. India and Bangladesh, they are coming from too many places, creating an ethnically super-diverse landscape in a city like London. Moreover, people used to permanently migrate to a host country. Today, they often have a kind hybrid status; thanks to cheap international flights and new communication technologies, people regularly travel to their home country and are seamlessly connected to their homeland, catching up with every single piece of family and political news. This has changed the idea and lifestyle of diaspora communities.

The hybrid status has also become very common among academics who are involved in international research collaborations, and highly skilled and specialised individuals working in international corporations engaged with joint-venture projects. Gradually, it has become the norm for people to live and work in two different countries. Personally, I was a researcher in different German institutes for some six years, while my family lived in London. Therefore, I was commuting between London and Germany. Initially, I assumed that this was not a normal lifestyle; however, I soon realised that I was not alone in this nomadic way of life. I met the same people on my regular flights, including a friend who was living in London and teaching in a German university. I had another friend who was engaged with a joint-venture project and commuting between London and Copenhagen every week; an Indian colleague in Germany whose family were in Amsterdam; and a German friend who lived in Germany but was a professor in Utrecht. Settling back in London, I got to know the head of a university department who lives in Italy, and another friend who teaches in London and takes the Eurostar every week to commute between London and a German city. The list goes on.

Globalisation denationalises economies, but immigration is one of the consequences of globalisation that renationalises politics (Sassen, 1996: 65). While the financial crisis of the late 2000s and the neoliberal economy are the backbone of the sociopolitical circumstances leading to Brexit, the Leave campaign focused on the immigration issue to effectively stimulate nationalist sentiment, blaming the 'other' for the mess. In other words, while the left-leaning scholars blame neoliberalism, austerity policies and lack of public spending for the recent sociopolitical frustration, the new far-right populist ideologues blame multiculturalism and immigration as an existential threat to Western culture and society, selling the dream of reviving the glorious past.

In the following discussion, I shall discuss the historical geography of globalisation and neoliberalism in a broad historical perspective to study the sociopolitical circumstances leading to the EU referendum. The historical trajectory of global geography advances our ability to comprehend different aspects of the geography of globalisation, enabling us to think about the post-Brexit geographies.

The Geography of Globalisation and Colonial Geography

Globalisation is about interconnecting production processes and trade across national-political borders. Globalisation is not only defined in very diverse ways, making it a polarised and contested concept, but it is also referred to in different historical periods. While globalisation is often considered a recent process that has changed our way of life in the late 20th century, many scholars argue that globalisation has centuries of history. Abu-Lughod (1991, 2005) has argued that contemporary globalisation is neither a new nor a unique historical phenomenon, rather it is the latest stage in the long historical process of global capitalist urbanisation. Therefore, she has insisted on the importance of a long-term historical perspective to grasp contemporary globalisation (Abu-Lughod, 2005: 53).

The early civilisations emerged in isolation in Mesopotamia, Egypt, today's Pakistan and China, then in the Mediterranean region and Meso-America. However, there is evidence that the three civilisations along the Nile, the Tigris-Euphrates and the Indus rivers interconnected by the first trans-regional exchange and trade system from the second millennium BCE (e.g. Pacione, 2005: 42–47). Cross-regional trade routes expanded over time and Abu-Lughod (1991) has argued that a trading system between the Middle East and China was well established by the 12th century. The establishment of Muslim rules and the Moghul Empire contributed to this interconnected geography, which is well captured by Ibn-Battuta's journey. This Moroccan scholar left his homeland to go on a hajj pilgrimage, but this famous journey lasted some 30 years during the 13th century. Ibn Battuta's travelogue titled *A Gift to Those Who Contemplate the Wonders of Cities and the Marvels of Traveling*, narrates a journey from Iraq, Persia, Arabia and Somalis, to the Swahili Coast, then from Anatolia, Crimea, Central Asia, South Asia, South East Asia to China; the last leg of his journey during 1249–1254 was from Spain, North Africa and Mali to Timbuktu. In this journey, he performed the hajj pilgrimage several times and served the Indian Royal court for almost a decade. Although this had been a unique journey, it captures the vast interconnected geography in which a scholar could freely travel, work and live.

Abu-Lughod has explained that the first world-system of trade was created when Europe joined the Eastern interconnected civilisations during the 12th century. This system was based on the three interdependent subsystems of Europe, the Middle East and the Far East that interacted on more or less equal terms. However, from the middle of the 14th century, the old world-system gradually evolved into the modern system which was dominated by the West. Nonetheless, it was the rise of Western colonial powers and their so-called 'discovery' of the New World, as well as the Dutch invention of capitalism that fundamentally evolved and shaped a global system of exchange and trades. This was the early phase of the trade system that Immanuel Wallerstein (1974) has called 'the modern world-system', a system that in geographical terms has truly reached a global scale. The modern system is characterised by the invention of capitalism, the rapid rise of European colonial powers and the expansion of the global system into the New World across the Atlantic. Wallerstein (1974, 1976) has argued that the 'world-system' simultaneously connects the world while reinforcing its inequalities through a single division of labour.

The modern world-system was not merely about expanding trade networks; it was the result of new technological innovations, political agendas and the development of capitalism, that reconfigured the global geography. The juxtaposition of capitalism, the Industrial Revolution and colonial imperialism evolved the mode of production, and reconfigured the global geography. Karl Marx (1848: 4) argued that capitalism naturally tends to form a world market in which capital freely moves without physical or political barriers. While new technical innovations remove physical barriers, the colonial system eliminated the political borders to create a new geography which was an ideal habitat for capitalism. The British Empire reshaped the global geography by its management of natural and human resources, production processes and consumption across the largest empire in human history. A part of the world was the source of raw materials, another territory was the agricultural zone to feed the Empire, while the heart of the Empire was an industrial hub. Human populations were dislocated, either as slaves or subjects, across the British colonies on an industrial scale, creating a new global division of labour. For example, Indian diaspora are spread in more than 135 countries, mostly relocated in Africa, the Caribbean and Oceania as a result of the emerging world capitalist system and the revolution in transportation and communication technologies (Dubey, 2003; Jayaram & Atal, 2004). The new global arrangement not only changed the perceived and lived geography, but it also created a new cognitive map of the world. While human and political geography expanded, the perception of the world shrank and the global geography became imaginable even for the average person. *Around the World in 80 Days* (Verne, 1873) not only well captures the truly globalised networks of communications and transportation, but also narrates the imagination of the global geography, produced by British colonialism.

The importance of an imagined geography is that although colonialism ended, its perception remains as an active part of (mis)understanding the global order, sustaining and reviving colonial mentality. Porter (2014: 393–394) has argued that one of the common after-effects of empires is that they leave behind after-images of a bright bulb; its shape can be seen for a while, if we close our eyes. For example, Iran was never colonised, although the British had extensive numbers of concessions in Iran.[3] Nonetheless, the colonial mentality is very vivid in Iranians' mind-set. In his memoir and more recent book, Jack Straw (2012, 2019) jokingly explained that Iranians are the only people in the world who believe that the British Empire still exists and rules the world. He argues that this mentality is rooted in the great influence of the UK in shaping the Iranian political landscape during the late 19th century, concession over the Iranian oil industry until the 1950s and British involvement in the 1953 US–UK coup against Prime Minister Mohammad Mossadegh, who nationalised the Iranian oil industry. The Iranian public still believe that the British are behind any important change in global order, and any major change would be defiantly in favour of the UK. To my amazement, my Iranian friends understood the result of the EU referendum based on this mentality. In fact, they shared this colonial mentality with the leaders of the Leave campaign. One of the arguments that Dorling and Tomlinson (2019) have made about politicians behind the Leave campaign is that their Brexit fantasy is based on a colonial mentality that was still dominant in British education in the 1960s and 1970s. Many of these politicians, often from rich and privileged families, went to schools that aimed at educating men to run an empire that no longer existed. In these elite schools, the colonial culture of superiority still defines their teaching culture. The result is that Boris Johnson argued that Britain is 'going to make a titanic success' out of Brexit.[4]

The history of colonial geography has another fold, that is the impact of the British Empire on its local political geography. The geography of the UK and the Empire have changed in a parallel but interlinked process. The colonial history contributed to constructing the Union and establishing solidarity among England, Scotland, Wales and Northern Ireland. These countries may have their own defined historical references: Death of the last Welsh prince (1282); the Battle of Bannockburn for the Scottish (1314); the Battle of the Boyne (1619) or/and the 1916 failed revolution for the Irish; and the battle of Waterloo and the defeat of Napoleon for the English (1815). Nonetheless, the history of Empire and then the two world wars have contributed to produce shared historical references. Although it is often argued that the British Empire had been dominant and was in favour of England, each of the four countries contributed to the British Empire and benefited from it. For example, it is hard to imagine the British Empire and capitalism without the contribution of Scotland to the Industrial Revolution.

What is important for the discussion in hand is that the political geography, that defining the union of four countries, has been very fluid and dynamic. For example, the Irish border was only recognised in 1921, when the British Empire was at its peak, controlling 24% of the globe. The Good Friday Agreement was only achieved in the 1990s after a long period of political tension and violence. This agreement was one of the most important political achievements in recent British political history.

The multiple geography of Northern Ireland contributed to implementing and sustaining this agreement. Geographically, Northern Ireland is located in the UK and the EU. This political geography makes it possible not only to diminish a hard border within the island of Ireland, but also creates a space in which people of Ireland produce a 'lived geography' in which the unification of Northern Ireland and the Republic of Ireland can be experienced in everyday life, without the existence of such a political geography.

Globalisation and Neoliberalism

The year 1979 is associated not only with the Iranian Revolution and my aforementioned déjà vu, but it is also the year when Margret Thatcher became prime minister, two events that signified a new era in the global order. Thatcher was at the forefront of promoting neoliberalism and deregulation of the market as the sole alternative in the new global landscape of the postcolonial era. The British Empire crumbled at an incredible rate during the 1960s, and almost all colonies with a significant population were decolonised by the late 1970s. In this new landscape, the welfare state was not an option; the government should have reduced its role and given privilege to the private sectors. In this circumstance, as a social Darwinists, she glorified inequality and argued that: 'It is our job to glory in inequality and see that talents and abilities are given vent and expression for the benefit of us all' (Wade, 2011: 54).

David Harvey, among other Marxist scholars, focuses on the periodical crisis of capitalism, suggesting that although Western capitalism acts like a clever magician who keeps showing an unexpected trick, these tricks will come to an end. Nonetheless, Western capitalism has historically shown an incredible resilience to survive throughout changing circumstances. Thatcher was probably right, evolve or die. Western capitalism that needed a global market was faced with a fundamental problem since the political geography of the world was profoundly changed in the postcolonial era. These changes were so profound that Platvoet (2004) argued that the end of the colonial era even contributed to the evolution of anthropology as an academic discipline, since many anthropologists were forced out of former colonies and had no choice but to study their own societies. The Western powers should invent a new global order to outlast their political and economic hegemony. This was gradually institutionalised following the end of the Second World War by (a) founding international institutions, such as the UN, the International Monetary Fund (IMF) and the World Bank – controlled by Western powers; and (b) promoting neoliberal ideology to deregulate the markets, giving privilege to Western international corporations. These developments ultimately created a new global order that we call globalisation. The process of globalisation reached its peak with the revolution in communications technology, ever cheaper and faster international shipping and transportation and, most importantly, the end of the Cold War. During the post-Second World War era, the global geography was configured based on Western–Eastern political geography, creating a major barrier/border to the free flow of labour, commodities and goods. The collapse of the Soviet Union and the disintegration of the Socialist/Eastern bloc in the 1990s accelerated the process of globalisation; it was then that globalisation captured the public discourse and imagination, a process that was often described as the Americanisation of the world. For example, Sassen (1996: 17–22) argued that one of the key features of globalisation is Americanisation since the global market is mainly regulated by a very few American private credit-rating agencies.

Globalisation is recognised by Abu-Lughod as the latest form of 'Western urban capitalism', in which there is almost no barrier to the flow of commodities and products; money/capital can seamlessly move between global market hubs. Urban hubs have played a great role in the global network of trade during both the medieval and modern times. Abu-Lughod (1991) explained the importance of cities as trade hubs for global networks of trade, when the world's population was not urbanised and predominantly settled in countries. She described trading cities as islands standing above a vast sea, creating a network in which the world economy was functioning. These cities were described by Peter Hall (1984) as 'world cities', such as Venice, Amsterdam and London, that channelled international forces and wealth towards their own national interest. Although globalisation is recognised by Abu-Lughod as the latest form of 'Western urban capitalism', the globalisation of the late 20th century has unique and very distinctive characteristics, most important of which is the emergence of 'global cities' as defined by Sassen.

New communication technologies and the internet have made trading seamless across the world, and a neoliberal ideology provides a platform to allow international corporations to dominate the world economy. Based on these technological developments and political ideology, the new global geography was configured based on 'global cities' such as London, New York and Tokyo. As mentioned, globalisation produced a unique geography that has a duality: while globalisation disperses the geography of production and consumption, its geography is heavily centralised around 'global cities'. These cities function as the command and control centres of the world economy. There is a fundamental difference between 'world cities', as described by Peter Hall, and 'global cities' articulated by Sassen (2001, 2005); she has argued that there is no such entity as a single global city, by definition. Rather there is a network of global cities, which contribute to growth in their own network (aka international corporations) and may not necessarily function in the interest of nations. Therefore, London as a part of the global cities network functions differently compared with London as a 'world city' during the colonial era.

London is a unique city in the UK, not because of its size and population, or its contribution to the UK's gross domestic product (GDP), but rather because of its position in the network of global cities. While Westminster has always been the seat of the British government, the city of London was reinvented as one of the most important global cities throughout the process of globalisation under neoliberal ideology. The reinvention of London as a global city financially *denationalised* and socioculturally *cosmopolitanised* this city, producing a sociopolitical landscape that is distinctively different compared with the rest of the country. The idea of London as a global city should not be reduced to the idea of financial or banking sectors, or the capitalist elite class. This city is a locale where political power, global wealth and cultural diversity are accumulated. The idea of London as a global city explains why London has a unique position in the EU referendum. Although London and Scotland voted to remain in the EU (59.93% and 62%, respectively), the sociopolitical logic behind their vote is very different. London is a hub of 'remainees' since it is a global city; and Scotland has a kind of national logic for its vote.

Globalisation has not only internationalised the market and denationalised global cities, but it has also produced supra-national or non-territorial solidarities. On the one hand, colonial capitalism, entangled with nation-states, evolved into an economic system that is associated with international corporations. On the other hand, social solidarities increasingly constructed around supra-territorial issues such as global warming and universal values such as lesbian, gay, bisexual, transgender's (LGBT) rights and the Occupy Movement against global capitalism (e.g. see Scholte, 2000). Today, even new nationalist movements are globalised, this is considered as a paradox in new far-right/nationalist movements. Although the recent far-right movements are against globalisation, because they are constructed and developed on globalised infrastructures, such as Facebook, globalisation is embedded in their DNA. Conventional nationalism aimed at creating solidarity among fragmented territories that were under feudal or local rules. However, the new nationalism is aimed at de-globalising the world geography in which Western countries are no longer dominant. The recent right-wing nationalist movements differ from conventional anti-globalisation movements associated with left ideology, such as the Occupy Movement. These new nationalists are aimed at neither building nation-states, nor dismantling Western global capitalism or neoliberalism. From the Brexit Party to the one-man party of Freedom in the Netherlands, they have no concrete manifesto or political solution for current crises. From Donald Trump and Boris Johnson to Nigel Farage and Geert Wilders, they all offer no actual alternative. What slogans such as *Make America Great Again* and *Make Britain Great Again* share is that they promise and offer the *feeling* of a glorious past and reviving of the lost supremacy.

The last aspect of the current global landscape that I would like to address is the condition of urbanisation, which plays an important role in shaping the future of the global order. The world population is increasingly urbanised and the centre of urbanisation has shifted to Asia and the Global South, where fast-growing megacities are shaping the future of an inter-connected global landscape of urbanisation. Today, the landscape of urbanisation is described as planetary urbanisation, addressing the fact that urbanisation processes affect the entire planet and the world's population, not just particular places known as urban areas. While only a century ago, 20% of the world's population lived in cities, this figure reached 53% in 2010 (Burdett & Sudjic, 2011; Ewen, 2016). Soon, the landscape of urbanisation will be dominated by megacities. While 26 cities were recognised as megacities with populations exceeding 10 million inhabitants in 2011, the UN estimates that there will be 41 megacities in 2030, the majority of which will be in the Global South.

Nation/modern states were the primary actors in negotiating and shaping world affairs and the global geography (e.g. see Gilpin, 1981: 18; Held, 2002: 16). In the second half of the 20th century, not only international political institutions, e.g. the UN, but also transnational corporations have gained an important position on the global stage. Soon, the social and political magnitude of megacities will make them one of the primary actors in shaping the future of global geography.

Conclusion

Here, geography is examined as a spatial system in which social, political, economic and technological developments are unfolded and evolved. The discussion throughout the chapter articulates how the idea of capitalism evolved in different political and technological landscapes. As argued, the geography may constitute multiple layers which are indistinguishably entwined. It is often socioeconomic, political or technological changes that make us aware of such complex geography. The chapter also discussed how imagined/cognitive geography is an active part of the historical dynamics of our geography.

The results of the EU referendum in 2016 have already made an impact on our perception of the global order. The public debates about Brexit have already revived the demand for a second Scottish independence referendum, highlighted the issue of the Irish border, boosted far-right movements, fractured the Labour and Conservative parties due to internal party disagreements over Brexit policies and made international corporations aware that the geography of the global market based on neoliberal ideology is crumbling. Boris Johnson moved into No 10 in June 2019 based on a promise that he would get the UK out of the EU to gain back sovereignty for the UK, and to deliver his promise he even tried to prorogue Parliament which breached the principle of parliamentary sovereignty.

Nobody knew the meaning of Brexit in 2016, and still nobody knows what the consequences of Johnson's political motions will be. It is difficult to predict the landscape post-Brexit. However, as the parliamentary negotiations on delivering Brexit fully exposed the complexity of the unwritten British constitution, delivering Brexit will certainly expose the anatomy of the multilayered geography of the UK. It will reveal that purging one layer of such a geography, i.e. membership of the EU, would even disintegrate the configuration of its interlaced layers. It is not unimaginable that a no-deal Brexit will lead to an independent Scotland, and even decolonisation of Northern Ireland, as Anglophobic Irish nationalists would describe it. Even if Johnson was ousted from No 10 by Parliament and Brexit was derailed, still the geography of the UK, and the UK's position in the global geography would not be as before; our cognitive map of the UK's geography is already transformed. Although there is no way to predict the outcome of this irreversible historic transitional process, we are certainly not going to walk back to the *glorious* geography of the colonial era or the late 20th century. This is the end of globalisation as we know it, but it is not the end of an interconnected world. A new global geography is in the making through a process in which not only nation-states of the developed world, international political bodies and global corporations, but also megacities of the Global South, will play a principle role. The question is: Will London be able to reinvent itself, remaining an important city in the global geography? Or the British will talk about the glorious past of this great city in the future.

Notes

(1) This ideology promotes a capitalist free-market, privatisation, the deregulation of markets, free trade, austerity and minimising government spending to increase the role of the private sector in the market and society.
(2) This is rephrased by different wording; it may have been popularised by Michael Harrington's (1962) book, *The Other America*.
(3) Colonialism was not entirely based on direct colonial rule; colonial power was fulfilled in different ways, such as concessions, leaseholds, protectorates, extraterritoriality and the mandate system (see Langsam, 1939: 19–21, 1962: 55–57).
(4) Obviously, there is irony in this phrase. When George Osborne among the audience shouted: 'it sank', Boris Johnson added: 'Well, the Titanic exhibition in Northern Ireland is the single most popular attraction in the province, [...] We are going to make a colossal success of Brexit' (Elgot, 2016).

References

Abu-Lughod, J. (1991) *Before European Hegemony: The World System A.D. 1250–1350*. New Edition. New York/Oxford: Oxford University Press.

Abu-Lughod, J. (2005) Global city formation in New York, Chicago and Los Angeles: An historical perspective. In N. Brenner and R. Keil (eds) *The Global Cities Reader* (pp. 42–48). New edition. London/New York: Routledge.

Burdett, R. and Sudjic, D. (eds) (2011) *Living in the Endless City*. London: Phaidon.

Curzon of Kedleston, G. (1915) *Subjects of the Day: Being a Selection of Speeches and Writings*. London: Allen & Unwin.

Dorling, D. and Tomlinson, S. (2019) *Rule Britannia: Brexit and the End of Empire*. London: Biteback Publishing.

Dubey, A. (ed.) (2003) *Indian Diaspora: Global Identity*. Delhi: Kalinga Publications.

Elgot, J. (2016) Brexit will be titanic success, says Boris Johnson. *The Guardian*, 3 November. See https://www.theguardian.com/politics/2016/nov/03/brexit-will-be-titanic-success-says-boris-johnson (accessed 7 August 2019).

Ewen, S. (2016) *What Is Urban History?* Cambridge: Polity.

Fukuyama, F. (1989) The end of history? *The National Interest* (16), 3–18.

Gilpin, R. (1981) *War and Change in World Politics*. Cambridge: Cambridge University Press.

Hall, P. (1984) *The World Cities* (3rd edn). London: Weidenfeld and Nicolson.

Harrington, M. (1962) *The Other America: Poverty in the United States*. New York: Macmillan.

Harvey, D. (2006) *Spaces of Global Capitalism: A Theory of Uneven Geographical Development*. London: Verso.

Held, D. (2002) *Globalization/Anti-Globalization*. Oxford: Polity Press.

Jayaram, N. and Atal, Y. (eds) (2004) *The Indian Diaspora: Dynamics of Migration*. (Themes in Indian Sociology v. 4). New Delhi: Sage Publications.

Langsam, W. (1939) *In Quest of Empire: The Problem of Colonies*. New York: Kessinger Publishing, LLC.

Langsam, W. (1962) Techniques of imperialism 1870–1939. In L. Snyder (ed.) *The Imperialism Reader: Documents and Readings on Modern Expansionism* (pp. 55–61). Port Washington/London: Van Nostrand.

Marx, K. (1848) *Manifesto of the Communist Party* (1888 edn. Pub 1908). New York: New York Labor News Co.

Pacione, M. (2005) *Urban Geography: A Global Perspective* (2nd edn). London: Routledge.

Platvoet, J. (2004) Ritual as war: On the need to de-Westernize the concept. In J. Kreinath, C. Hartunge and A. Deschner (eds) *The Dynamics of Changing Rituals: The Transformation of Religious Rituals within Their Social and Cultural Context* (pp. 243–266). New York: Peter Lang.

Porter, B. (2014) Epilogue: After-image of Empire. In K. Nicolaidis, B. Sèbe and G. Maas (eds) *Echoes of Empire: Memory, Identity and the Legacy of Imperialism* (pp. 391–406). London/New York: I.B. Tauris.

Said, E. (1979) *Orientalism*. New York: Vintage Books.

Sassen, S. (1996) *Losing Control? Sovereignty in an Age of Globalization*. New York: Columbia University Press.

Sassen, S. (2001) *The Global City: New York, London, Tokyo*. Princeton, NJ: Princeton University Press.

Sassen, S. (2005) Locating cities on global circuits. In N. Brenner and R. Keil (eds) *The Global Cities Reader* (pp. 89–95) (New edition). London/New York: Routledge.

Scholte, J. (2000) *Globalization: A Critical Introduction*. Basingstoke: Palgrave.

Straw, J. (2012) *Last Man Standing: Memoirs of a Political Survivor*. London: Pan.

Straw, J. (2019) *The English Job: Understanding Iran and Why It Distrusts Britain*. London: Biteback Publishing.

Verne, J. (1873) *Around the World in Eighty Days*. Paris: Gauthier-Villars.

Vertovec, S. (2007) Super-diversity and its implications. *Ethnic and Racial Studies* 30 (6), 1024–1054.

Wade, R. (2011) Global trends in income inequality. *Challenge* 54 (5), 54–75.
Wallerstein, I. (1974) *The Modern World-System. Studies in Social Discontinuity.* New York/London: Academic Press.
Wallerstein, I. (1976) *The Modern World-System: Capitalist Agriculture and the Origins of the European World-Economy in the Sixteenth Century* (text edition). Studies in Social Discontinuity. New York: Academic Press.
Young, R. (2001) *Postcolonialism: An Historical Introduction.* Oxford: Blackwell Publishers.

Coda 2020: COVID-19 Masks but Fails to Flatten Brexit

Hazel Andrews

Brexit's Fall

After David Cameron, then prime minister of the UK, announced in February 2016 that there would be a referendum on UK membership of the European Union (EU), media outlets were dominated by reports, speculation and debates as to the possible implications of exiting and what way the vote would go. This intensified once the result was known, with questions being asked about when Article 50 (the legal mechanism by which a country gives notice to leave the EU) would be triggered, what the withdrawal agreement would look like, what was the actual leaving date and so on. Some of this speculation ended with the replacement of Cameron's successor Theresa May by Boris Johnson as prime minister in July 2019 and the Conservative Party's landslide victory in the December 2019 general election. With a decidedly pro-Brexit cabinet now running the country, the date of departure was no longer uncertain, and the UK officially left the EU on 31 January 2020. What was to follow was the negotiation of a new relationship between the two entities. As I write in June 2020, these negotiations are ongoing, but they seem to be rarely reported, or at least, if they are, have much less of the prominence that they once had. Brexit has given way to COVID-19.

COVID-19 is the name given to the latest strain of a novel coronavirus, the third of its kind this century. It was first identified in the city of Wuhan, Hubei Province, China, in December 2019. By January 2020, the highly infectious virus had spread to 24 countries. The disease continued to spread in February and by the second week of March, the World Health Organisation (WHO) had declared a pandemic. As the situation continued to develop, and before the WHO proclamation, many countries had implemented airport screening of arrivals and instigated border controls restricting travel to and from China, although these measures were later demonstrated to be of limited impact in arresting the virus's spread (Wells et al., 2020). Within countries, governments began to enact measures of 'lockdown' demanding that their citizens stay at home, only being permitted to leave their place of abode for essential items or for work that could not be undertaken at home. Different countries did this at different times and at different rates. In the UK, lockdown was implemented on 23 March 2020. Prior to this, the need for social distancing and hand sanitation was already being rehearsed in numerous public settings and the general public had already begun to prepare for restrictions with the stockpiling of food and sanitary products – hand sanitiser, toilet paper – already taking place. The stay-at-home order in the UK meant that for many people they lost income, as they could either no longer work or they lost a proportion of their earnings. In addition, people were prevented from seeing family and friends in person. With such fundamental changes to practices of daily life, attention, of course, shifted to the more immediate matter in hand. It is, therefore, not surprising that the current, ongoing UK–EU negotiations on a new relationship are not front page news.

Nevertheless, the UK remains in transition and Brexit rumbles on. Albeit obscured from view, debate continues as to Brexit's merits and drawbacks and whether the UK's final departure from the EU will be with a deal or with no deal. Further, questions remain as to whether the talks to resolve this question need to continue beyond the end of the 2020 transition deadline, because due to COVID-19 fewer meetings have been held than if the pandemic had not occurred. Delays to the negotiations to make up for lost time are not the only reason to extend discussion; in addition, the economic hardship that the catastrophe of the pandemic brings could potentially be partially ameliorated by preventing further disruption to the national economy that changes to existing relationships on either path of a new deal or a no deal will inevitably bring.

As I point out in the introduction to this book, the full ramifications for tourism and Brexit cannot yet be fully articulated because they are hitherto unseen, and because of COVID-19 they may never be completely revealed. However, the collection of essays in this book was never intended to be an impact assessment of Brexit on tourism, although its implications may at times be discussed. Rather, it has been to open ideas of what a coalescence of Brexit and tourism tells us about the nature of the sociocultural world in which these things are practised, especially in relation to travel, borders and identity. However, COVID-19 perhaps detracts from the clarity provided by this lens as it has displaced Brexit from the media spotlight and conceivably many people's minds. So, how to make sense of Brexit when COVID-19 is thrown into the mix? Both represent unchartered waters being 'unprecedented' situations: no country has left the EU before and, despite the 1918 Spanish Flu pandemic, COVID-19 offers its own hitherto unencountered challenges. I heed well Catherine Palmer's warning in her contribution to this book (Chapter 3), that we should be wary of making a thesis when in the 'eye of the storm'; but rather we should wait for clearer skies to assess outcomes. What follows, then, are reflections on issues relating to travel, borders and identity in light of COVID-19. These reflections tentatively draw out some of the interlacing of political reactions and attitudes towards the virus and its handling with those relating to Brexit given that both pose major ruptures in the sociocultural, as well as the economic, fabric of the UK.

The proceeding discussion is necessarily from a perspective limited by lockdown and social distancing. It begins with a short discussion to contextualise the use of a stay-at-home approach.

From My Armchair

The idea of 'armchair anthropology' is most associated with the discipline's earliest days and embryonic beginnings. Some of the main theorists of this time, notably James Frazer (1854–1941) and Edward Burnett Tylor (1832–1917), have been criticised for not directly engaging with the places and peoples that they wrote about, rather they relied on information gathered by others. The change in approach, as is well known, was heralded by Bronisław Malinowski whose emphasis on collecting data first-hand by undertaking 'fieldwork' in 'the field' laid the foundations for modern ethnographies. As Berger (1993: 174) attests '"fieldwork" is – it goes without saying, and thus must be said – the sine qua non of modern anthropology, the ritual initiation experience (Freilich 1970: ix) in the discipline, and the fundamental characteristic distinguishing modern anthropology from its pre-Malinowskian existence – "armchair anthropology"'. However, Efram Sera-Shriar (2014) argues that the apparent rupture between being in an armchair and not being in an armchair is a distortion, because the pre- and post-armchair ways of working have more synergies than are often recognised. Sera-Shriar (2014: 27) claims 'armchair anthropology was not a passive pursuit, with minimal analytical reflection that simply synthesized the materials of other writers'. During the time of lockdown and the requirement to stay at home and maintain social distancing, going out to conduct research is problematic; as such, my 'field' of enquiry, upon which my reflections are based, is therefore, necessarily, based on the world viewed from my armchair. Another caveat I wish to make relates to also not making a claim to be offering a holistic presentation of the current circumstances, which in any event are in constant flux, and so in this instance I follow James Urry's assertion about ethnography, that it 'does not attempt to "record" the totality of everyday life in a particular context' (Urry, 1993: 2 in Sera-Shriar, 2014: 32). In the same vein, I do not claim to be recording every instance of relevance to lockdown UK.

Project Fear

During the 2016 referendum debates, the supporters of Vote Leave dismissed the predicted damage to the UK of leaving the EU as 'Project Fear' and touted, in some quarters, as the remain side's best chance of winning the vote.[1] One 'fear' to be highlighted was the loss of freedom of movement and the ramifications for touristic practices. The coronavirus has brought restrictions on the freedom of movement centre stage, not just for the UK but globally. In April 2020, the United Nations World Tourism Organisation noted that of 217 destinations worldwide, 72% had stopped international tourism.[2] In addition, in the first quarter of 2020, international tourism was down by 22% with predictions in May 2020 for a potential decline of between 60% and 80% over the whole year.[3]

COVID-19 understandably invites fear because of the risk of either becoming seriously ill, or dying, if one were to catch the virus. As the disease started to take a foothold in the UK, common courtesies and everyday intimacies such as greeting someone with a handshake or kiss on the cheek were forsaken by many, as Manderson and Levine (2020: 367) comment in relation to South Africa, '[we have seen a] retraction of social engagement, shifting from the joking "Are we still kissing?" To maintaining "social distance" without remark'. However, as if to almost echo the idea of what is all the fuss about if we leave the EU, because it will make 'us' (including the National Health Service) stronger,[4] Boris Johnson was filmed shaking hands with COVID-19 patients and then openly boasting about it during a live address to the nation in which he proclaimed, 'I shook hands with everybody',[5] as if to allay fears that the unfolding situation was really that serious; but perhaps also mirroring views expressed by ordinary citizens elsewhere in the world from Pakistan to Hungary, that there was nothing to fear (Ali, 2020: 2).

However, despite Johnson's bravado, people were scared. Writing from their perspective of being situated in the Global South, Manderson and Levine (2020: 367) argue the fear that people were feeling is based not just on the threat of infection, but also on awareness of the toll the global pandemic could take in relation to social and economic life. They note the historical evidence that highlights how fears of contagion fuel social exclusion and discrimination. As Delan Devakumar *et al.* (2020: 1194) writing in the *Lancet* observe, 'throughout history, infectious diseases have been associated with othering', and the coronavirus pandemic has opened the door for xenophobia and racism. They argue that some political leaders have used the virus to 'reinforce racial discrimination' (Devakumar *et al.*, 2020). Devakumar *et al.* do not specifically call out the UK in this respect; nevertheless, there have been cases of racially aggravated COVID-19 attacks on people, particularly of Asian descent, in the country (Coates, 2020). Fear of the other was fuel to the Vote Leave campaign, which highlighted concerns about Turkey joining the EU as a potential threat to the UK, a danger that would, it was claimed, be ameliorated by 'taking back control' from the EU of national borders and being able to prevent the other from entering the UK. Of course, viruses, like fish, have no knowledge of, or concern for, national borders. It seems that both COVID-19 and Brexit have acted as a catalyst for racially motivated crimes, as following the result of the referendum there was a spike in such cases in the UK.

A very English affair?

Boris Johnson's somewhat laissez-faire attitude to the arrival of the coronavirus in the UK meant that according to many commentators the implementation of policies to curb the virus's spread was slow.[6] Some of this reluctance has been attributed to Johnson's desire to focus on Brexit and the EU negotiations, to continue to play on his popularity by marking 31 January 2020 with grand public celebrations. In addition, there was some advice to allow the virus to take its course to facilitate 'herd immunity', as well as the idea that if lockdown were imposed at the wrong time, the populace would too soon get fatigued by it and therefore not adhere to the rules.[7] Timings may well come back to haunt Johnson, not only in relation to the imposition of lockdown, but also in its lifting (to which I will return) and also in his determination to persist with Brexit by ending the transition period on 31 December 2020, regardless of the results of the negotiations with the EU and irrespective of the economic ramifications of a global pandemic.

The reference to 'herd immunity' places the virus in the world of nature, but as Manderson and Levine (2020: 367) argue, COVID-19 is very much part of culture: 'the institution of quarantine practices, lockdowns, and border controls, and the insistence on adherence to hygiene practices (handwashing) highlight how human practices and behaviors are implicated, and foreshadow a global humanitarian crisis as community transmission takes hold'. Perhaps Johnson had misjudged the culture he believed he so readily appealed to, for as Manderson and Levine (2020: 368) point out, while noting that the ability to comply with the rules relates to class, race and gender, 'people can change behaviour quickly when presented with institutional mandates to do so' and, from their perspective of medical anthropology, 'embodied knowledge seems to be more flexible than often theorised'. By the time the prime minister had mandated lockdown on 23 March 2020, the public had already begun to take matters into their own hands, stockpiling food and toilet paper, cancelling non-essential meetings, starting to wear face masks and clearing out stocks of hand sanitisers from the shops. Stockpiling had also been a characteristic of Brexit for some. Those fearful of reductions in food supplies by the implementation of border controls were hoarding at home[8] and, in fact, the government had indulged in its own hoarding in the form of stockpiling some medicines. It is stating the obvious that lockdown restricts travel and therefore has implications for the practice of tourism, reducing it on a global scale. The lack of freedom of movement in response to COVID-19 and Brexit is another thing that both issues have in common, although in the case of the former it is much more pronounced.

When Johnson addressed the nation to impose lockdown, he referred to a 'national character' that would not like restrictions to its freedom of movement (although, ironically, this is one of the results of Brexit). Announcing lockdown, he stated 'I know how difficult this is, how it seems to go against the freedom-loving instincts of the British people'.[9] Journalist Fintan O'Toole later picked up on the retelling of Johnson's words in *The Sun* newspaper which they interpreted as 'Mr Johnson said he realised it went against what he called "the inalienable free-born right of people born in England to go to the pub"'. O'Toole notes the exclusion of the Scots, Welsh and Northern Irish from this version, as well as those who live in the UK but were born elsewhere, and argues that *The Sun*'s interpretation makes more sense as, 'what Johnson was really evoking was a very specific English sense of exceptionalism, a fantasy of personal freedom as a marker of ethnic and national identity'.[10] This links well with the suggestion that Brexit is very much an English affair. As the UK moves to finalise its departure from the EU, all the entities that make up the country will leave, although not all the devolved parties returned a leave majority. Thus, despite the UK having the word 'united' in its name, this does not necessarily find fruition in practice. Following the referendum, analysis of voting patterns highlighted differences in voting based on region, class and education. COVID-19 has also highlighted a fractured, rather than a 'united' kingdom. Attempts to bring a national sense of unity could be found on 8 May 2020 when the country was invited to take part in social-distanced celebrations to mark the 75th anniversary of VE Day. Fortunate timing for Johnson, as he could once again resort to the rhetoric of war to promote a sense of national identity, albeit a mainly English one. According to *Guardian* newspaper columnist John Harris, 'Britain was led into the disaster of Brexit by people successfully sowing the ludicrous idea that subjecting ourselves to self-harm would somehow awaken a Blitz spirit and revive past glories. Amid Friday's juxtaposition of the 75th anniversary of VE day and a deepening sense of national crisis, as well as solemn remembrance, there was inevitably some of the same stuff. These things play into deep elements of the English psyche'.[11]

We're not all in it together

The idea at a time of such deep crisis is that, as Johnson's announcement in March 2020 implied, 'we are all in this together'. Behind the rules he invoked were ideas that everyone is potentially at threat from being infected and/or passing on the disease, so we must all play our part and observe the lockdown rules, making sacrifices, as necessary, along the way – families and friends separated, incomes diminished, jobs lost and travel plans cancelled. For the village of Eyam, in the Derbyshire Dales, 2020 was not the first time that a request to protect lives by foregoing certain freedoms had been made. During the years of the Great Plague of 1665 and 1666 in which parts of England, but notably London, were experiencing an outbreak of bubonic plague, Eyam went into voluntary isolation to prevent the spread of the disease. Thought to have been imported into the hamlet by the local tailor receiving material infested by plague-infected fleas, the disease spread through the village. The local Reverend – William Mompesson – managed to persuade most village residents to stay within the boundaries of the village, despite the obvious difficulties and dangers associated with so doing. Many of the village's wealthier residents, however, departed before lockdown began. Mompesson's own children also left, but he himself, having instigated the rules abided by them and remained with his wife in the village, although she later succumbed to the plague and died.[12]

The inequalities echo down the ages as the impacts of COVID-19 disproportionally impact members of the community based on ethnicity, class, age and gender not only in terms of susceptibility and likely death from the virus, but in the economic ramifications of lockdown.[13] In addition, social distancing rules have not been universally upheld, including among politicians and government advisers. Research undertaken by Ben Ansell (2020: 15) has shown that in broad terms those most likely to break the social distancing guidelines are more likely to have voted for Brexit than those who voted to stay within the EU, 'the Brexit divide that has shaped much of British life over the past half-decade is indeed showing up here'. Ansell's research is complex and considers several factors. His findings come with a number of caveats, for example someone identified as low-skilled and in the category of more likely to have voted for Brexit may also be a key worker, who is unable to implement the same measures of stay at home as someone more able to work from home and falls into the category of more likely to live in a remain area of the country. Ansell (2020: 15) attests that it is not the case that people did not social distance at all, but rather there are 'differences in the degree of social distancing'. Nevertheless, he argues that political attitudes are a strong factor shaping the underlying trends and 'in both the poorest and richest places, the more Remain-voting areas have distanced more' (Ansell, 2020: 9). A similar difference emerged when lockdown restrictions were partially lifted in England. A survey by the Office of National Statistics[14] revealed that 47% of leave supporters were in favour of the government's easing of lockdown compared with approval from Remainers being at 26%.[15]

It is probably the lessening of the restrictions, more than the actual restrictions, that has highlighted the differences, because different parts of the country have moved at different paces. The relaxation of lockdown rules announced by Boris Johnson on 10 May 2020 applied only to England. The devolved governments of Wales, Northern Ireland and Scotland maintained the 'stay-at-home' message.

'I didn't know Wales isn't in England'

The varying approaches to lockdown in the UK have thrown into sharp relief the nation's internal borders. I live in Wales, I can see the border to England from where I live so my examples are drawn from here, although there are doubtless similar scenarios to be found on the Scottish–English border. The lifting of lockdown restrictions in England caused quite a bit of consternation for many who live on or near the boundary. For example, that people in England could go and play a round of golf, whereas those who live in Wales could not brought the Welsh–English border into focus in a way that may not have been thought about before as crossing points rarely warrant much attention. However, for the village of Llamymynech, which straddles both Montgomeryshire/Powys, Wales, and Shropshire, England, the border suddenly became a big issue. The village's golf course is also located partly in Wales and partly in England, with most of the holes in the former. The club references its cross-boundaryness on their website, declaring 'The course claim to fame is that you can play one round of golf in two separate countries, courtesy of Offa's Dyke running through our course. On the 4th hole you tee off in Wales and putt out in England'.[16] Further north in Flintshire the town of Saltney is also split between England and Wales with some residents on the aptly named Boundary Lane becoming acutely aware of the different countries in which they resided. As one local, who lives on the English side, proclaimed: 'I went for a walk and it occurred to me that I was crossing over into the Welsh and then back over into England – normally I wouldn't think about it. I was thinking technically whether I should be entering Wales'. Further, a local councillor from Welsh Saltney spoke of the confusion people had expressed in comments and questions to her about the situation while at the same time noting that 'people thought "they were in the UK" and would have liked to have seen a more united front as lockdown restrictions are eased'.[17]

Perhaps the most outstanding event to draw attention to the different entities of Wales and England is the widely reported video clip of a woman who had driven 160 kilometres from her home in the English midlands to Wales to visit the Barmouth beach in Gwynedd. The video showed her remonstrating with the police, who had advised her that despite what Boris Johnson had told those living in England, the same rules did not apply in Wales and she was not permitted to stay on the beach. In a later posting on her own social media, the woman admitted she had been wrong and according to news reports she attributed her mistake to not being aware that Wales was not in England.[18]

Beach fever or 'I must go down to the seas again'[19]

Not wishing to have her desire for sea and sand thwarted by different lockdown rules, the woman who claimed not to know that Wales is not in England indicated her intentions to visit an English beach the next day, advising in her post 'But tomorrow we going to an English beach. F*** you Wales you bunch of b*******'.[20] It seems that she is not alone in her desire to journey outside of her immediate environment and head for the coast. Once restrictions were partially lifted in England, many of the country's beaches were packed with visitors. Indeed, one beach – Botany Bay on the Kent coast – was described as being, 'as busy as Notting Hill carnival' with large groups of people gathering during the day and some remaining to camp overnight. A similar scenario was repeated at other beaches around the English coast, as was an increased number of visitors to the Peak District.[21]

It is obvious that touristic practices will be impacted by COVID-19, because of the restrictions on the key ingredient – the ability to travel – that allows both national and international tourism to flourish. What tourism will look like in a post-lockdown world is yet to be determined. Some may remain fearful and restrict their own freedom of movement. The impact of Brexit on touristic practice both for inbound and outbound tourists may be harder to establish as disaggregating the ramifications of Brexit from COVID-19 will likely be difficult to clarify. Nevertheless, people want to travel. Before the global pandemic, the numbers of international tourist arrivals were in the billions. These numbers may well return, even considering the events of 2020. Indeed, despite high profile media reporting of cruise ships as 'ideal incubators of infectious diseases',[22] bookings in some parts of the world for cruises in 2021 show an increase.[23] As borders begin to open and tourist destinations once again receive tourists, many social distancing practices will remain in place along with the use of face masks where people are in closer proximity such as in an airplane cabin or a restaurant.[24] It is to the use of face coverings that I now wish to turn to draw this afterword to a close.

The masking of Brexit

The wearing of face coverings by Muslim women in both France and the UK has caused controversy. In the former, laws were introduced that prohibited the wearing of full face coverings in 2011. In the UK, former Labour government Home Secretary (1997–2001) Jack Straw caused controversy with his suggestion that the wearing of the veil created divisions within communities (Eli, 2020). More recently (2018), writing a column for the UK newspaper *The Daily Telegraph*, Boris Johnson described women in burkhas as looking like 'letter boxes'. His comments have been attributed to an increase in Islamophobia.[25] With such debates it seems ironic, then, as Eli (2020: 742) points out, that 'the piece of cloth that once caused so much aggravations – particularly as a security threat – mainly in European countries seems to be nowadays an essential and mandatory part of human daily precautions in the face of the coronavirus pandemic'. It is the case that in the UK the voluntary wearing of face coverings in public from across the sociocultural spectrum has risen and is now mandatory on the London transport network. The wearing of face coverings is an attempt to curtail the virus's spread and is becoming a part of accepted daily practice. At the same time, COVID-19 is masking Brexit, not just in that the latter has fallen from the spotlight of media attention but that the ramifications of Brexit in terms of tourism and other sectors of the economy are likely to be obscured as the full impact of the UK's departure from the EU will be harder to disentangle from the fallout from the coronavirus, as the disease will doubtless become a covering of ever greater density for any negativity associated with Brexit. The coronavirus may have displaced Brexit from headline UK news, but beneath the mask it continues apace. Thus, unlike the government's apparent desire to flatten the virus rate of infection curve, COVID-19 has not flattened Brexit.

According to Igor Calzada (2020), the pandemic has slowed the rhetoric of globalisation and the ideal of a borderless world. However, there are commentators who argue that it is precisely a global approach that is required to effectively exit from the lockdown limitations. For example, Ngaire Woods and Rajaie Batniji (2020) call for increased cooperation across national borders to restart the economy, ensure adequate testing and tracing of the virus and the distribution of a vaccine if/when one becomes available. They state that it is 'only by ditching nationalist rhetoric and policies, and embracing stronger international cooperation, can governments protect the people they claim to represent'. Even Boris Johnson, a key figure responsible for Brexit and the erecting of boundaries between the UK and her continental European neighbours, has called for global cooperation against COVID-19.[26]

The global pandemic and the UK's departure from the EU are quite different phenomena. However, the responses by governments and individuals to both give us a lens through which questions of identity and belonging can be examined. Both occurrences also give insight into the importance of travel, and responses to its restrictions. So much is unknown and cannot be divined about the future of the UK following 31 January 2020, and any such divination will be clouded by the global pandemic. However, as the country continues to travel, despite COVID-19, in a new direction, with its departure from the EU the new order of things will materialise in cultural practices of which both travel and tourism are expressions. Border watching has never seemed more important.

Notes

(1) https://www.newstatesman.com/politics/uk/2016/06/project-fear-back-and-its-still-remains-best-hope (accessed 4 June 2020).
(2) https://www.unwto.org/news/COVID-19-world-tourism-remains-at-a-standstill-as-100-of-countries-impose-restrictions-on-travel (accessed 8 June 2020).

(3) https://www.unwto.org/news/COVID-19-international-tourist-numbers-could-fall-60-80-in-2020 (accessed 8 June 2020).
(4) http://www.voteleavetakecontrol.org/statement_by_michael_gove_boris_johnson_and_gisela_stuart_for_the_sun_vote_leave_to_cut_vat_on_fuel.html (accessed 4 June 2020).
(5) http://www.voteleavetakecontrol.org/statement_by_michael_gove_boris_johnson_and_gisela_stuart_for_the_sun_vote_leave_to_cut_vat_on_fuel.html (accessed 4 June 2020).
(6) It should be noted that the Westminster-based government is only responsible for the English part of the UK and that the leaders of the devolved governments of Wales, Scotland and Northern Ireland are responsible for these countries as separate entities.
(7) https://www.theguardian.com/world/2020/apr/18/how-did-britain-get-its-response-to-coronavirus-so-wrong (accessed 4 June 2020).

https://www.newstatesman.com/politics/uk/2020/03/why-aren-t-we-lockdown-coronavirus-yet (accessed 4 June 2020).
(8) https://www.theguardian.com/lifeandstyle/2019/dec/10/brexit-stockpilers-still-have-enough-for-three-months (accessed 4 June 2020).
(9) https://www.gov.uk/government/speeches/pm-statement-on-coronavirus-20-march-2020 (accessed 21 March 2020).
(10) https://www.theguardian.com/commentisfree/2020/apr/11/coronavirus-exposed-myth-british-exceptionalism (accessed 13 May 2020).
(11) https://www.theguardian.com/commentisfree/2020/may/11/bunting-britain-covid-19-crisis-nationalist (accessed 27 May 2020).
(12) https://www.derbytelegraph.co.uk/news/health/eyam-derbyshire-bubonic-plague-quarantine-3948082 (accessed 27 May 2020).
(13) https://news.sky.com/story/coronavirus-were-all-in-this-together-but-some-more-than-others-11981917 (accessed 5 June 2020).
(14) https://docs.cdn.yougov.com/oygox62yaw/YouGov%20Covid%20Lockdown%20Results%20May%202020.pdf (accessed 26 May 2020).
(15) https://www.theguardian.com/commentisfree/2020/may/20/nudge-theory-brexit-divide-lockdown-coronavirus#maincontent (accessed 26 May 2020).
(16) https://www.llanymynechgolfclub.co.uk/ (accessed 8 June 2020).
(17) https://www.dailypost.co.uk/news/north-wales-news/north-wales-road-two-different-18238826 (accessed 5 June 2020).
(18) https://www.independent.co.uk/news/uk/home-news/coronavirus-lockdown-wales-gwynedd-beach-england-boris-johnson-a9531271.html (accessed 3 June 2020).
(19) The first phrase of John Masefield's (1878–1967) poem *Sea Fever* published in 1902.
(20) https://www.independent.co.uk/news/uk/home-news/coronavirus-lockdown-wales-gwynedd-beach-england-boris-johnson-a9531271.html (accessed 3 June 2020).
(21) https://www.theguardian.com/uk-news/2020/may/25/busy-as-notting-hill-carnival-botany-bay-residents-bemoan-packed-kent-cove (accessed 27 February 2020).
(22) https://www.theguardian.com/commentisfree/2020/apr/14/cruise-ships-coronavirus-passengers-future (accessed 21 April 2020).
(23) https://nypost.com/2020/04/01/cruise-bookings-are-on-the-rise-for-2021-despite-coronavirus/ (accessed 21 April 2020).

https://www.seatrade-cruise.com/news/singapore-positions-support-cruise-tourism-revival?fbclid=IwAR3SUh2Hemo0rf_QTaNteikDBphtTlxUuBOisYdZ_La85lO1OSdM6v9vn3s (accessed 21 April 2020).

https://qz.com/1830415/despite-coronavirus-outbreaks-cruise-bookings-are-up-for-2021/ (accessed 21 April 2020).
(24) https://www.theguardian.com/travel/2020/jun/03/its-an-exciting-beginning-venice-opens-to-tourists (accessed 11 June 2020).

(25) https://www.pressgazette.co.uk/boris-johnson-telegraph-column-muslim-women-letterboxes-bank-robbers-spike-islamophobic-incidents/ (accessed 9 June 2020).
(26) https://metro.co.uk/video/boris-johnson-calls-global-cooperation-eu-led-coronavirus-summit-2164874/ (accessed 9 June 2020).

References

Ali, I. (2020) The COVID-19 pandemic: Making sense of rumor and fear. *Medical Anthropology*. doi.org/10.1080/01459740.2020.1745481.

Ansell, B. (2020) What explains differences in social distancing in the UK? See https://ukandeu.ac.uk/wp-content/uploads/2020/04/What-explains-differences-in-social-distancing-in-the-UK.pdf (accessed 13 May 2020).

Berger, R.A. (1993) From text to (field)work and back again: Theorizing a post(modern)-ethnography. *Anthropological Quarterly* 66 (4), 174–186.

Calzada, I. (2020) Will Covid-19 be the end of the global citizen? Apolitical. See https://apolitical.co/en/solution_article/will-covid-19-be-the-end-of-the-global-citizen doi: 10.13140/RG.2.2.11942.27208/1 (accessed 11 June 2020).

Coates, M. (2020) Covid-19 and rise of racism. The BMJ letter published 6 April 2020. See http://www.bmj.com/ (accessed 12 May 2020).

Devakumar, D., Shannon, G., Bhopal, S.S. and Abubakar, I. (2020) Racism and discrimination in COVID-19 responses. *The Lancet* 395, 1194.

Eli, T.B. (2020) The anthropology of the face mask: Rethinking the history of face covering controversies, bans and COVID-19 context. *Journal of Xi'an University of Architecture & Technology* X11 (V), 741–751.

Manderson, L. and Levine, S. (2020) COVID-19, risk, fear, and fall-out. *Medical Anthropology* 39, 367–370. doi.org/10.1080/01459740.2020.1746301.

Sera-Shriar, E. (2014) What is armchair anthropology? Observational practices in 19th-century British human sciences. *History of the Human Sciences* 27 (2), 26–40.

Urry, J. (1993) *Before Social Anthropology: Essays on the History of British Anthropology*. Chur: Harwood Academic.

Wells, C.R., Sah, P., Moghadas, S.M., Pandey, A., Shoukat, A., Wang, Y., Wang, Z., Meyer L.A., Singer, B.H. and Galvani, A.P. (2020) Impact of international travel and border control measures on the global spread of the novel 2019 coronavirus outbreak. *PNAS* 117 (13). See www.pnas.org/cgi/doi/10.1073/pnas.2002616117 (accessed 27 May 2020).

Woods, N. and Batniji, R. (2020) A global COVID-19 exit strategy. See https://www.project-syndicate.org/commentary/covid19-exit-strategy-requires-global-cooperation-by-ngaire-woods-and-rajaie-batniji-2020-04 (accessed 26 May 2020).

Index

Act of Union, 36
Africa, xvi, 64, 125, 140, 143-4, 193-4, 206
African, xvi-xviii,76, 125, 127, 131, 133, 135-7, 139-40, 154, 165
African-American, 76
Age, 8, 46, 51, 82, 86, 97, 104, 107, 175, 209
Alliance of Small and Island States (AOSIS), 159
American, 15-16, 44, 64, 67, 71, 73, 76-7, 140, 145, 165, 173, 197
Americanisation, 197
Anglo-cluster, 144
Anglophone, xvii-svii,13, 135, 142, 152
Anglo-Saxon, 112
Anglosphere, 141, 144-6, 152-5
Anthropology, xiii, xxi, 5, 16-17, 35, 112, 123, 196, 205, 207, 214
Anthropology of Tourism, Travel & Pilgrimage (ATTP), xiii, xvi-xvii, xx-xxi
Anti-EU, 99, 102, 105
Anti-Europe, 104
Anti-free movement, 176
ANZUS pact, 143
Article 50 of the Treaty on European Union, 10, 47, 188, 203
Asia, 44, 81, 142, 155, 171, 173, 193, 199
Asian, 19, 143, 155, 129, 206
Associate EU Citizenship, xv, 14, 174-87
Association of Commonwealth Universities, 76
Asylum, 27, 191
Austerity, xix, 8, 104, 191-2, 200
Australia, 28, 66-7, 71-4, 77-9, 142-4, 146, 150-3, 155, 161
Authenticity, 55, 60, 103-4, 107, 149
Autonomy, 75, 112, 120, 166, 170

Belfast, 50
Benefit tourism, 27, 183
Benidorm, xiv, xvii,12, 96-109
Biodiversity and Ecosystem Services in Territories of Europe Overseas, 165
Blair, Tony, 2, 69, 110-11
Blitz, 71, 147, 208
Bombay, 151
Brexit Britain, xiv, xvii-xix
Brexiteers, xv
Brexit Party, 111-12, 143, 198
Brexit-speak, xv, xviii
Britishness, 12, 34- 36, 38-9, 41, 43, 46, 61, 77, 80, 89, 93, 98, 101, 106-7, 113-15, 117, 123, 129-30, 146, 148, 153, 157, 162
Brussels, xv-xvi, 46, 111, 172, 186
Bureaucracy, 83, 115, 116

Cameron, David 6, 49, 57, 61, 63, 99, 108, 110, 203
Canada, 28, 31, 71, 77, 74, 142-4, 146, 152
Capitalism, 2, 4, 190-1, 193-9
Cardiff, 41, 50
Caribbean, 47, 77, 89, 143-5, 149, 161, 165, 168, 194
CARICOM, 165, 172-3
CARIFO-RUM, 165
Case 26/62 Van Gen den Loos, 29
Case 6/64 Costa v ENEL, 29
Case C-184/99 Grzelczyk, 181
Celtic, 48-65, 144
Central and Eastern Europe, xvii-xviii, 12, 80-95
China, 85, 146, 190-1, 193, 203
Chinese, 130-1, 138, 142
Christian-Democrat ideology, 20

Churchill, 28, 30, 110, 146-7, 154
Citizens' Rights Directive (CRD), 181-2, 184
Class, xiv, 2, 8, 12, 37, 52, 64, 68, 75, 96, 98-101, 103-5, 107-8, 115, 127-8, 131, 150, 182, 190-1, 198, 207-9
Cold War, 146, 189-90, 197
Colonialism, 13, 42, 50, 61, 70, 90, 112, 124-7, 131, 133, 135, 139, 142, 144-5, 147, 149-55, 158, 160-2, 171, 189, 190, 193-8, 200
Commodification, xv, 55-6, 62, 138, 149
Common Market, 23, 28, 31-2, 67, 107-08, 186-7
Commonwealth, xvi-xxi,11-13, 28-9, 38, 67, 70-1, 73-9, 141-56, 160, 166, 172, 189
Commonwealth Games, 76, 142
Commonwealth of Nations, 70, 141-2
Commonwealth Trust, 76
Communitas, 162, 170
Community, 3, 5, 6, 11, 18, 21- 2, 29, 31, 37, 40, 55, 63, 72, 96, 115, 147-8, 152, 159, 161, 165, 167-8, 190, 207, 209
Conservative Party, xv-xix, 2, 6, 8, 9-10, 14, 49, 98-9, 102, 105-8, 110, 123, 199, 203
Constitution, 43, 111, 145, 188, 200
Consumerism, xv, 126, 136
Consumption, 189
Continental Europe, 11, 120, 138
Control, 5, 9, 13, 44-5, 94, 103, 108, 110-24, 141, 158, 175-7, 191, 197, 214
Core–periphery, 13, 157-73
Cornwall, 12, 48-65, 75, 77
Cosmopolitan, xiv, 20, 104, 198
Council of Europe, 21, 32
Court of Justice of the European Union (CJEU), 176-8, 180-3
COVID-19, 11, 39, 203-14
Cruise ships, 165, 167, 169, 211
Cuba, 77
Culture, xvii, 3, 13, 16, 23, 26-7, 42, 45-6, 53-4, 60, 87, 89, 99-100, 105, 115, 117, 141, 143, 145, 155-6, 168, 181, 192, 195, 207
Customs Union, 19, 22

Decolonisation, 200
Democracy, 23, 33, 38, 142, 145, 152, 154, 185, 190
Denationalised, 192, 198
Department for International Development (DFID), 167, 172
Dependency, 160, 162, 166, 171
Dependent Territory, 162
Deregulation, 196, 200
Deutschland Erwake (Germany Awake), xvii
Devolution, 50
Diaspora, 81, 135, 172, 192, 194
Disneyfication, 55, 57
Displacement, xvi-xix
Division of labour, 194
Dutch, 83, 128-9, 132, 164, 168, 193
Duty Free, 12, 96-109

European Convention on Human Rights (ECHR), 178
Edinburgh, 41, 50, 70
Education, 23, 27, 32, 37, 53, 60, 88, 97, 104, 132, 134, 150-1, 164, 195, 208
EEC, 22-3, 98, 103, 106, 115, 143-4, 152
Elitism, 35, 38, 60, 68, 103, 112, 146, 150-2, 171-2, 190-1, 195, 198
Embodiment, 86, 94, 100, 117, 139, 207
Emigration, 83, 87, 104
Empire, xiv, 2, 19, 28-9, 38, 49, 70-3, 78-9, 101-2, 106-7, 112, 116, 141-56, 151,158, 162, 167, 193-6, 201
Enemies of the People, 111
England, 8-9, 36, 38, 42, 45, 478, 50-1, 53, 57, 61, 63, 66, 88-90, 92, 94, 98, 106, 112, 115, 117, 120, 145, 153, 195, 208-11
English, 8-9, 16, 36, 38, 41, 46, 48, 50-51, 53, 57, 59- 61, 63, 78, 87-90, 92, 94, 98-100, 105, 112, 114, 121, 129-36, 144-6, 148, 153, 155, 195, 201, 207-08, 210-11, 213
English Heritage, 50- 51, 53, 57, 59- 60, 63, 90
English Reformation, 50
Englishness, 38, 45-6, 50, 64, 94
Ethnicity, 9, 24, 36-7, 45, 54, 61, 81, 83, 85, 88-9, 114, 141, 149-9, 152, 191-2, 208-9

Ethno-centrism, 103
Ethnography, xiv, xvi-xvii, xx, 12-13, 38, 46, 48, 64, 94, 113, 205, 214
EU Citizens, 28, 30, 104, 175, 178, 180-1, 183-4, 187
EU Citizenship, xvii, 14, 19, 174-87
EU Commission, 178, 186
EU Select Committee, 163-4
EU Withdrawal Bill, xviii
EU Workers, 28, 102
Europa, 19- 20, 32, 140
European Capital of Culture, 3, 81
European Coal and Steel Community (ECSC), 22, 32
European Commission, 3, 23, 26-7, 30-1, 81, 166, 172
European Consciousness, xv, 11, 18, 25
European Convention on Human Rights, 178
European Court of Justice, 29
European Development Fund (EDF), 163-6
European Economic Community, 22, 66, 106, 116, 143
European elections, 111
European Football Championship, 83
European Frame-work Convention for the Protection of National Minorities, 50
European integration, 3, 11, 18-28, 30, 32, 116
European Parliament, 176
European Song Contest, 3
European Spirit, 21, 32
European Union Charter of Fundamental Rights, 30
European unity, 11, 20-1, 23, 27, 30, 32
Europeanness, 3
Eurosceptic, 49, 64, 74, 108, 174
Euroscepticism, xvi,29
Exceptionalism, 184, 208, 213
Exotic, 100, 105
Expatriate, 10, 96, 100, 115

Face masks, 207, 211
Falkland Islands, 143, 161-2, 173
Farage, Nigel, xviii, 99, 105, 111, 143, 198
Far-right, 7, 15, 27, 192, 198-9

Fear, 2, 7, 15, 104, 106, 134, 139, 205-7, 211-12, 214
Federal, 22, 29, 153
Fieldwork, 8, 12-13, 38, 48-51, 62, 113, 127-8, 205
Fiji, 77, 142, 147-8, 150-1
First World War, 70
Five Eyes intelligence alliance, 143
Five Power Defence Arrangements, 143
Foreign Office, 75
Founding Fathers, 18, 20-21, 25, 27, 29
Fragmentation, xiii, xviii-xx
France, 22, 29, 36, 50, 67, 81, 83, 85-6, 107, 116, 161, 164, 211
Free movement, xv,14, 19, 23, 25, 27, 30, 104, 110-124, 143, 146, 174-187, 180, 183, 187
Free trade, 143-4, 146, 152, 200
Freedom, 1, 5, 9, 13, 16, 22-3, 31, 73, 101, 110-24, 136, 146, 154, 163, 181, 184, 186-7, 205-9, 211
Free-market, 190, 200
French, 20, 22, 32, 36, 75, 110, 130, 132-3, 135-6, 140, 150, 164, 166, 181
Fundamental Rights, 174-87
Fundamental Status, 174-87

Gaelic, 50
Gallos, 57-8, 60
Gender, 37, 44-5, 45, 51, 108, 116 207, 209
Gentrification, 91, 93, 136
Geoffrey of Monmouth, 52, 57
Geopolitical, 11, 34, 128, 135, 160, 162-3, 169, 170-1, 173
German, 3, 13, 32, 70, 82-3, 110, 113-14, 125-40 158, 192
Germanness, 129, 139
Germany, 22, 29, 32, 36, 46, 78, 82-3, 85-7, 91, 114, 116, 126-7, 151, 182, 192
Gibraltar, 161-4
Glastonbury, 12, 51
Global cities, 191, 197-8
Global market, 196-7, 199
Global order, 141, 189-91, 194-6, 199
Global South, xvi, 125, 136, 191, 190-200, 206

Globalisation, 14, 89, 115, 170, 188-202, 212
Good Friday Agreement, 195
Great Britain, 34, 36, 38, 41-2, 66-7, 74, 77-8, 82, 88, 90, 92, 94, 108, 144

Habitus, 5, 99
Hadrian's Wall, 41, 53
Harare Declaration, 142
Hard Brexit, 19, 45, 158
Hegemony, 12, 21, 54, 145, 190, 196
Heritage, 2, 12-13, 15, 26, 30-2, 41, 42, 45, 48-65, 84, 88-90, 92, 141, 146-9, 151, 153-4, 167
Historic Buildings and Monuments Commission, 53
Home Office, xvii
Homophobic, 99
Hospitality, xx-xxi, 4-5, 15, 86
Hostile, xix-xx 50, 131-2, 134, 149
Hostile Environment, xvii, xx
House of Commons, 28, 31, 156
House of Lords, 29, 163, 186

Ideology, 131, 135, 140, 190, 196-200
Image, 2-3, 6, 16, 30, 58, 77, 81, 83, 85-6, 88, 91-3, 97, 100, 102-3, 110, 116, 122, 125, 127, 168, 194, 201
Imaginaries, 6, 31, 34, 88, 97, 135, 155
Imagination, xix, 1, 4-5, 19, 22, 48, 51, 54, 86, 89, 94, 97, 101-2, 105, 118, 141, 161, 171, 194-5, 197, 199
Immigrants, xvii-xviii, xxii, 7, 14, 83, 90, 101-2, 135, 140, 145, 147, 149, 167, 191
Immigration, 1, 7, 9, 27, 29, 49, 51, 64, 102, 111, 176, 189, 191-2
Imperialism, xiv, 15, 33, 45, 70-4, 90, 107, 112, 114, 126, 146, 148, 172, 190, 194, 201
Inauthentic, 68
Independence, 22, 29, 38, 41, 44, 116, 142, 177
India, xiv, xxii, 56, 72-3, 107, 115, 120, 129, 142-5, 152, 192
Indian, xiv, 47, 71-2, 89, 131, 138, 142, 147, 161, 192-4, 201
Industrial Revolution, 194-5
Inequality, 51, 116, 158, 181, 196, 202

Interculturalism, xv
International Monetary Fund (IMF), 196
International Tourist Year, 24
International Year of Peace, 25
Internationalisation, 89
Iran, 14, 188, 194-6, 201
Ireland, 8, 34, 36, 41, 50, 85, 91, 142, 144-5, 189, 196
Irish, xiv, 7, 9, 39, 46-7, 50, 85, 189, 195, 199-200, 208
Irish border, 7, 189, 195, 199
Iron Curtain, 80, 91, 146, 154
Islam, 3, 49
Islamic Republic, 14, 188
Islamophobia, 155, 211

Jamaica, 117
Johnson, Boris, 7- 8, 43-4, 110, 123, 143, 188, 195, 198-203, 206-212
Joint Ministerial Council on European Negotiations (JMC-OT EN), 163

King Arthur, 48-65

Labour Party, 2, 8-10, 49, 98, 106, 110, 199, 211
Labour Red Wall, 8
Language, 7, 11, 27, 39, 43, 49-50, 54, 60, 85, 87-8, 92, 100-1, 105, 112, 114, 121-2, 125-41, 144-6, 162
Latin Americans, 129
Leave campaign, xiv, 75, 101-3, 107, 116, 120, 190, 192, 195, 206
Leave UK, 117
Leave voters, 104
Leavers, 7-8, 11, 45
Liberal Democrat, 8
Liberty, 112, 121-4, 145
Liminality, xiii- xxii, 1-17, 127, 129, 162, 170, 173
Lived experience, 36, 97, 190
Lived geography, 194, 196
Lockdown, 203-14
London, 15-17, 21, 31-2, 41, 44-7, 49-50, 60, 62-4, 71, 78-9, 84-5, 88, 90, 94-5, 107-8, 123-4, 140, 145, 147, 151, 153-6, 171-3, 186, 188, 191-2, 197-98, 200-2, 209, 212
Low-cost airlines, 80, 91, 105

Index 219

Maastricht Treaty, 3, 23, 175
Magaluf, xiv, 13, 96, 100, 110-24, 126-32, 135, 137-39
Magic, xvi, 1-17
Magician, 196
Magna Carta, 145
Malinowski, Bronisław, xx, 7, 16, 112, 123, 205
Mallorca, 8, 13, 105, 109-140, 146
Manila Declaration, 4, 24
Māori, 149, 151, 155
Marx, Karl, 194, 201
Mass tourism, 13, 25, 96-7, 108, 127
Mauss, Marcel, 5, 16
May, Theresa, xvii, 7, 76, 102, 108, 188, 203
Media, xiv, xix-xx, 6, 9, 12, 28, 30, 34, 38, 40, 46, 68-9, 76-7, 80, 83, 85, 88, 92, 97, 104-5, 114, 136, 145, 147, 167-9, 183, 186, 203-4, 210-12
Mediterranean, xviii, 52, 96, 108, 113, 126-7, 193
Megacities, 199-200
Meghan Markle, xiv, 12, 68, 76-8, 147, 153
Megxit, 76
MEPs, 111-12, 176-7
Metropolitan, 13, 159, 161, 167
Migrants, xv-xx, 13, 102, 108, 117, 126-140
Migration, 16, 19, 23, 28, 31, 38, 81, 84, 92, 102, 107, 115-16, 121, 127-8, 132, 135, 171, 174
Military, 142-3, 145-6, 162
Mobility, 11, 13-14, 18, 23, 25, 80, 103, 105, 139, 157, 176, 179-80, 182-5, 187
Monarchy, 41, 60, 67-9, 71, 73-4, 78, 147-8, 162
Moral panic, 102, 106
Morocco, 77, 103, 131
Multiculturalism, xiii, 12, 49, 62, 88, 90, 192
Muslim, 134, 149, 155, 193, 211
Myth, xiv, xix, 5-7, 19, 23-5, 31-2, 39, 56, 61, 64, 77, 118, 121, 150, 213

Napoleon, 36, 121-2, 195

Nation, 3, 15, 18-19, 21, 30, 34-47, 54, 63-5, 69-70, 74, 79, 84, 97, 110, 116, 127-8, 140, 148, 154, 157-8, 165, 183-4, 189-91, 198, 200, 206, 208, 210, 214
National Health Service, 40, 206
National identity, 34, 116
Nationalism, 2, 15, 20, 27, 38, 44, 50, 61, 90, 125, 127, 137, 162, 170, 192, 198, 212-13
Nationality, 10, 12, 93, 98, 113-14, 116, 123, 140, 175, 177, 181
Nation-ness, 37, 39-40
Nation-state, 54, 116, 157-8, 189
NATO, 75
Negative freedom, 117, 121
Neoliberalism, xv, 2, 98, 102, 106, 190-2, 196-9
Netherlands, 67, 85, 161, 178, 198
New World, 124, 193, 194
New Zealand, 66-79, 142-56, 161, 168
New Zealand Ministry for Culture and Heritage, 74, 79
NGOs, 146, 168
Nigeria, 131, 135, 140, 142
Nigerian, 129, 131, 135-7, 139
No-deal Brexit, 39-40, 46, 200, 204
Non-territorial, 198
North Atlantic Treaty Organization, 143
Northern European, 96
Northern Ireland, 8, 36, 41, 46-7, 50, 153, 189, 195-6, 200, 209, 213
North–south divide, 8, 15
Nostalgia, xiv, 77, 90, 141, 144, 147-8, 150-1, 155

Oceania, 144, 194
Overseas Countries and Territories (OCTs), 13, 145, 157-73
Othering, 100, 105-6, 206
Otherness, 98-9, 101, 103, 105, 161
Overseas Association Decision (OAD), 163-4

Package holiday, 28, 96-7, 105
Palmanova, 13, 110-24

220 Index

Pandemic, 11, 39, 203-14
Parliament, xviii, 8, 10, 26, 29, 31, 35, 41-4, 46, 49-50, 78, 90, 111-12, 123, 176, 180, 186-8, 200
Partnership for Progress and Prosperity, 162
Patriotism, 162
Peace, xv, 4, 9, 11, 18, 20-5, 27, 29-32, 34, 73, 121, 154
Pilgrimage, xiii, xxii, 193
Pitcairn Island, xvi, xxi,13, 157-73
Pluralism, 179, 184
Poland, xv-xvi,78, 80-95
Polish Tourist Organisation (PTO), 80, 85-6, 93
Political geography, 189, 194-7
Politicians, 7, 35, 45, 73, 80, 99, 106, 142, 146, 163, 190-1, 195, 209
Politics, 2, 15, 20, 27, 29, 36, 45, 61, 63-4, 70, 78-9, 90, 98, 105, 107, 111, 122-3, 127, 129, 141, 144, 153, 158, 160, 163, 171, 183, 187, 192, 201, 212-13
Populist, 27, 92, 108, 190, 192
Post-Brexit, xiii-xv, xvii, 10-12, 14-15, 39, 41-2, 44-5, 67, 75, 77-8, 91, 102, 121, 138, 141, 148, 152, 157-8, 162-4, 166, 169, 171, 173, 175-7, 185, 187, 192, 200
Postcards, 6, 58
Postcolonial, 90, 135, 190
Post-devolution, 12
Post-industrial, 2, 99, 108
Post-socialist, 81
Power, 13, 15, 35, 40, 43-4, 50, 54-5, 57-8, 69, 73, 75, 116, 141, 145-6, 148, 151-2, 157-9, 161, 163, 168, 170-1, 189-90, 198, 200
Prayer Book Rebellion, 50
Present-centeredness, 12, 58
Prince Harry, xiv,12, 68, 76, 147, 148, 153
Privy Council, 145
Pro-Brexit, 141, 144, 203
Pro-Europe, 104, 174
Project Fear, 7, 205

Race, xiv, xvi 2,12, 37, 75-6, 99-100, 102-3, 106-8, 125,135-7,139-40, 148-50, 152, 155, 206-7, 214

Referendum, 1-2, 4, 6-10, 13-17, 28-9, 35, 37, 44, 48-50, 61-3, 69, 90, 92, 99, 104, 108, 110-11, 132, 157-8, 162-3, 175, 185, 188-9, 192, 195, 198-9, 203, 205, 207-8
Refugee, xv, 27, 182, 191
Region, xx, 8, 15, 34, 37-8, 41-2, 44, 50-1, 57, 61-2, 64-5, 81, 86, 88-9, 91-2, 99-100, 123,140, 155, 169-70, 193, 208
Re-imagination, xvi, 141-156, 169
Religion, 20, 36-7, 54, 88, 148
Remainers, 7, 8, 11, 198, 209
Renationalisation, 192
Republic of Ireland, 8, 196
Republicanism, 69
Republicans, 36
Right-wing, 9, 106, 191, 198
Rites of Passage, xiii
Ritual, xiv, 5, 7, 10, 66, 68, 71, 78, 153, 205
Royal, xiv, xvii, 12, 16, 46, 52, 63, 66-79, 88-9, 110, 142, 144, 147-8, 150, 152-6, 168, 171, 173, 193

S'Arenal, 113, 125-140
Schengen, 82, 126
Schuman, Robert, xiv, 18-33
Scotland, 8, 16, 36, 41, 46, 50, 88, 93, 153, 195, 198, 200, 209, 213
Scottish, xiv, 8, 36, 38, 41, 50, 89-90, 114, 131, 195, 199, 208, 210
Scottish independence, 8, 50
Scottish National Party, 8
Second World War, 2-3, 12, 20, 34, 36, 71, 73, 83, 105, 116, 142, 145, 147-8, 196-7
Self-determination, 35, 112, 157, 162, 165, 167
Senegalese, 128-9, 132-7, 140
Sexuality, 37, 98-9, 101-2, 123, 127, 129, 138, 167
Shakespeare, 41, 90
Single Market, 18-19, 23, 26
Small Island Driven Economies (SITES), 159
Slavery, 42, 63, 112, 122, 151, 194
Subnational Island Jurisdictions (SNIJs), 158-60, 166

Social distancing, 204-5, 209, 211, 214
Social justice, 48, 55, 62-3
Social values, 175
Socioeconomic, 24, 26, 29, 98, 104, 128-9, 133, 184-5, 199
Socio-legal, 14, 180
Sociopolitical, 190, 192, 198
Solidarity, 18-20, 22-3, 29-30, 40, 157, 165, 183, 195, 198
South East Asian Treaty organisation, 143
South Pacific, 142, 169, 173
Souvenirs, xvi-xvii, 58-9, 74, 103, 115, 126, 129, 133, 138, 169
Sovereignty, 3, 22, 29, 35, 40, 42-3, 66, 69, 71, 74, 110-11, 116, 157-160, 189, 190, 199, 200
Spain, 3, 10, 12-13, 38, 63, 81, 85, 87, 96-140, 161, 193
Spanish, xiv, 6, 10, 13, 15, 96, 98-110, 113, 117, 130-5, 138-9, 146, 153, 204
Special Relationship, 75
Spiritual, 11, 18-33
Sri Lanka, 74, 147
St George's Day, 90
Sterling, 114
Stonehenge, 51, 53, 90
Straw, Jack, 195, 211
Subnational Island Jurisdictions (SIDS), 158, 173
Subregion, 81
Supranational, 3, 21-3, 25, 29
Sustainability, 26, 32, 63, 81, 91, 142, 160, 165-8, 172-3
Swedish, 123, 129
Symbolic, xiii-xv, 3, 6, 14, 32, 34, 37, 41, 54, 71, 74, 76, 85, 101-2, 116, 142, 147, 149, 151-2, 179
Symbolic violence, 6

Take Back Control, xv, 2, 9, 13, 101, 110-12, 116, 120-2, 141, 170, 206
Television, 66, 77, 89, 97-9, 145, 147
Terrorists, 75, 134
Thatcher, Margaret, xv, xix-xx, 98, 103, 106-7, 117, 196
The Commonwealth Secretariat, 142
The Crown, 36, 71, 74, 156
TIAC, 25
Tintagel, 12, 48-65

Tonga, 77
Torres Strait Island, 151
Tory Party, 29
Totalitarian, xix
Tourism development, 4, 13, 86, 91, 113, 160, 167-8, 170
Tourism industry, 4, 13, 15, 17, 25-6, 81, 126, 129, 138, 141, 149, 164-5, 169
Tourism Industry Association of Canada, 25
Tourists, 2, 6, 12-13, 24, 28, 34, 38-9, 41-2, 44, 51, 55, 61-2, 81-3, 85-8, 96-140, 141-56, 168, 211, 213
Trade, 13, 26, 28, 40, 62, 67, 75-6, 111, 141-3, 145, 147-8, 150-3, 155-8, 163, 165, 168-9, 189, 193-4, 197
Tradition, xiv, 2-3, 6, 9, 12, 21,23,27, 30, 48, 57, 60, 66, 68, 72, 77-9, 88-90, 103, 110, 108, 145, 150, 161, 181
Transition, 5, 10, 16, 39, 204, 207
Transnational, 185, 191, 199
Treaty of Rome, 22
Treaty on the European Union (TEU), 10, 20, 47, 175, 188, 203
Treaty on the Functioning of the European Union (TFEU), 6, 19, 25-6, 175, 180-1
Trieste, 3, 15
Trooping of the Colour, 68
Trump, Donald, 2, 170, 198
Turkey, 7, 206

UK Independence Party (UKIP), 75, 99-101, 106, 108-9, 111, 143
UK-Overseas Territories (UKOTs) 157-173
UK-Overseas Territories Joint Ministerial Council on European Negotiations, 163
Underclass, 75
UNESCO, 20, 32, 51, 91, 146, 155, 168
Union Jack, 105, 110, 114
United Nations, 18, 24, 51, 91, 159, 191, 196, 199, 206
United States, xvii, xix,12, 28, 73, 75-6, 78, 83, 85, 143-8, 152, 161, 169, 201
Urbanisation, 193, 199

Vaken Project, xvii, xxii

Valorisation, 55-6, 58, 83
Verhofstadt, Guy 176-7, 180, 185, 187
VisitBritain, 40-3, 47, 53, 65, 97, 95
VisitScotland, 41
Vote Leave, 1-2, 7-9, 13, 44, 110-11, 116, 120-2, 205-6

Wales, 2, 8, 36, 41, 50, 68-9, 71, 88, 92, 119, 147, 153, 195, 209-11, 213
Welfare state, 8, 196
Welsh, 36, 50, 89-90, 114, 195, 208, 210
West African, 13, 127-8, 133, 135, 137, 139-40
Western Europe, 3, 22, 86, 96

Western urban capitalism, 197
Westminster, 46, 49, 61, 145, 198, 213
Will of the people, xix
Windrush, xvii, 149
Withdrawal agreement, 175, 203
World Bank, 196
World Health Organisation, 203
World Heritage, 2, 51, 91, 168
World Tourism Organisation, 4, 17-18, 24-5, 28, 33, 80-1, 94, 206
World Travel Market, 84
World Youth Day, 81
World-system, 193-4

Xenophobia, 2, 17, 206

For Product Safety Concerns and Information please contact our EU Authorised Representative:

Easy Access System Europe

Mustamäe tee 50

10621 Tallinn

Estonia

gpsr.requests@easproject.com